104
Advances in Biochemical Engineering/Biotechnology

Series Editor: T. Scheper

Editorial Board:
W. Babel · I. Endo · S.-O. Enfors · A. Fiechter · M. Hoare · W.-S. Hu
B. Mattiasson · J. Nielsen · H. Sahm · K. Schügerl · G. Stephanopoulos
U. von Stockar · G. T. Tsao · C. Wandrey · J.-J. Zhong

Advances in Biochemical Engineering/Biotechnology
Series Editor: T. Scheper

Recently Published and Forthcoming Volumes

White Biotechnology
Volume Editors: Ulber, R., Sell, D.
Vol. 105, 2007

Analytics of Protein-DNA Interactions
Volume Editor: Seitz, H.
Vol. 104, 2007

Tissue Engineering II
Basics of Tissue Engineering and Tissue Applications
Volume Editors: Lee, K., Kaplan, D.
Vol. 103, 2007

Tissue Engineering I
Scaffold Systems for Tissue Engineering
Volume Editors: Lee, K., Kaplan, D.
Vol. 102, 2006

Cell Culture Engineering
Volume Editor: Hu, W.-S.
Vol. 101, 2006

Biotechnology for the Future
Volume Editor: Nielsen, J.
Vol. 100, 2005

Gene Therapy and Gene Delivery Systems
Volume Editors: Schaffer, D. V., Zhou, W.
Vol. 99, 2005

Sterile Filtration
Volume Editor: Jornitz, M. W.
Vol. 98, 2006

Marine Biotechnology II
Volume Editors: Le Gal, Y., Ulber, R.
Vol. 97, 2005

Marine Biotechnology I
Volume Editors: Le Gal, Y., Ulber, R.
Vol. 96, 2005

Microscopy Techniques
Volume Editor: Rietdorf, J.
Vol. 95, 2005

Regenerative Medicine II
Clinical and Preclinical Applications
Volume Editor: Yannas, I. V.
Vol. 94, 2005

Regenerative Medicine I
Theories, Models and Methods
Volume Editor: Yannas, I. V.
Vol. 93, 2005

Technology Transfer in Biotechnology
Volume Editor: Kragl, U.
Vol. 92, 2005

Recent Progress of Biochemical and Biomedical Engineering in Japan II
Volume Editor: Kobayashi, T.
Vol. 91, 2004

Recent Progress of Biochemical and Biomedical Engineering in Japan I
Volume Editor: Kobayashi, T.
Vol. 90, 2004

Physiological Stress Responses in Bioprocesses
Volume Editor: Enfors, S.-O.
Vol. 89, 2004

Molecular Biotechnology of Fungal β-Lactam Antibiotics and Related Peptide Synthetases
Volume Editor: Brakhage, A.
Vol. 88, 2004

Biomanufacturing
Volume Editor: Zhong, J.-J.
Vol. 87, 2004

New Trends and Developments in Biochemical Engineering
Vol. 86, 2004

Analytics of Protein–DNA Interactions

Volume Editor: Harald Seitz

With contributions by
C. Backendorf · V. Beier · M. L. Bulyk · S. Field
D. F. Fischer · J. D. Hoheisel · H. Lehrach · J. Majka
C. Mund · E. Nordhoff · J. Ragoussis · C. Speck · I. Udalova

Advances in Biochemical Engineering/Biotechnology reviews actual trends in modern biotechnology. Its aim is to cover all aspects of this interdisciplinary technology where knowledge, methods and expertise are required for chemistry, biochemistry, micro-biology, genetics, chemical engineering and computer science. Special volumes are dedicated to selected topics which focus on new biotechnological products and new processes for their synthesis and purification. They give the state-of-the-art of a topic in a comprehensive way thus being a valuable source for the next 3–5 years. It also discusses new discoveries and applications. Special volumes are edited by well known guest editors who invite reputed authors for the review articles in their volumes.

In references *Advances in Biochemical Engineering/Biotechnology* is abbeviated *Adv Biochem Engin/Biotechnol* and is cited as a journal.

Springer WWW home page: springer.com
Visit the ABE content at springerlink.com

Library of Congress Control Number: 2006934731

ISSN 0724-6145
ISBN-13 978-3-540-48147-8 Springer Berlin Heidelberg New York
DOI 10.1007/978-3-540-48150-8

This work is subject to copyright. All rights are reserved, whether the whole or part of the material is concerned, specifically the rights of translation, reprinting, reuse of illustrations, recitation, broadcasting, reproduction on microfilm or in any other way, and storage in data banks. Duplication of this publication or parts thereof is permitted only under the provisions of the German Copyright Law of September 9, 1965, in its current version, and permission for use must always be obtained from Springer. Violations are liable for prosecution under the German Copyright Law.

Springer is a part of Springer Science+Business Media

springer.com

© Springer-Verlag Berlin Heidelberg 2007

The use of registered names, trademarks, etc. in this publication does not imply, even in the absence of a specific statement, that such names are exempt from the relevant protective laws and regulations and therefore free for general use.

Cover design: WMXDesign GmbH, Heidelberg
Typesetting and Production: LE-TEX Jelonek, Schmidt & Vöckler GbR, Leipzig

Printed on acid-free paper   02/3141 YL – 5 4 3 2 1 0

Series Editor

Prof. Dr. T. Scheper

Institute of Technical Chemistry
University of Hannover
Callinstraße 3
30167 Hannover, Germany
scheper@iftc.uni-hannover.de

Volume Editor

Dr. Harald Seitz

Department Vertabrate Genomics
Prof. Dr. Hans Lehrach
Max Planck Institute for Molecular Genetics
Ihnestraße 63–73
14195 Berlin
seitz@molgen.mpg.de

Editorial Board

Prof. Dr. W. Babel

Section of Environmental Microbiology
Leipzig-Halle GmbH
Permoserstraße 15
04318 Leipzig, Germany
babel@umb.ufz.de

Prof. Dr. S.-O. Enfors

Department of Biochemistry and
Biotechnology
Royal Institute of Technology
Teknikringen 34,
100 44 Stockholm, Sweden
enfors@biotech.kth.se

Prof. Dr. M. Hoare

Department of Biochemical Engineering
University College London
Torrington Place
London, WC1E 7JE, UK
m.hoare@ucl.ac.uk

Prof. Dr. I. Endo

Saitama Industrial Technology Center
3-12-18, Kamiaoki Kawaguchi-shi
Saitama, 333-0844, Japan
a1102091@pref.saitama.lg.jp

Prof. Dr. A. Fiechter

Institute of Biotechnology
Eidgenössische Technische Hochschule
ETH-Hönggerberg
8093 Zürich, Switzerland
ae.fiechter@bluewin.ch

Prof. Dr. W.-S. Hu

Chemical Engineering
and Materials Science
University of Minnesota
421 Washington Avenue SE
Minneapolis, MN 55455-0132, USA
wshu@cems.umn.edu

Prof. Dr. B. Mattiasson

Department of Biotechnology
Chemical Center, Lund University
P.O. Box 124, 221 00 Lund, Sweden
bo.mattiasson@biotek.lu.se

Prof. Dr. H. Sahm

Institute of Biotechnolgy
Forschungszentrum Jülich GmbH
52425 Jülich, Germany
h.sahm@fz-juelich.de

Prof. Dr. G. Stephanopoulos

Department of Chemical Engineering
Massachusetts Institute of Technology
Cambridge, MA 02139-4307, USA
gregstep@mit.edu

Prof. Dr. G. T. Tsao

Professor Emeritus
Purdue University
West Lafayette, IN 47907, USA
tsaogt@ecn.purdue.edu
tsaogt2@yahoo.com

Prof. Dr. J.-J. Zhong

Bio-Building #3-311
College of Life Science & Biotechnology
Key Laboratory of Microbial Metabolism
Ministry of Education
Shanghai Jiao Tong University
800 Dong-Chuan Road
Minhang, Shanghai 200240, China
jjzhong@sjtu.edu.cn

Prof. Dr. J. Nielsen

Center for Process Biotechnology
Technical University of Denmark
Building 223
2800 Lyngby, Denmark
jn@biocentrum.dtu.dk

Prof. Dr. K. Schügerl

Institute of Technical Chemistry
University of Hannover, Callinstraße 3
30167 Hannover, Germany
schuegerl@iftc.uni-hannover.de

Prof. Dr. U. von Stockar

Laboratoire de Génie Chimique et
Biologique (LGCB), Départment de Chimie
Swiss Federal Institute
of Technology Lausanne
1015 Lausanne, Switzerland
urs.vonstockar@epfl.ch

Prof. Dr. C. Wandrey

Institute of Biotechnology
Forschungszentrum Jülich GmbH
52425 Jülich, Germany
c.wandrey@fz-juelich.de

Advances in Biochemical Engineering/Biotechnology Also Available Electronically

For all customers who have a standing order to Advances in Biochemical Engineering/Biotechnology, we offer the electronic version via SpringerLink free of charge. Please contact your librarian who can receive a password or free access to the full articles by registering at:

springerlink.com

If you do not have a subscription, you can still view the tables of contents of the volumes and the abstract of each article by going to the SpringerLink Homepage, clicking on "Browse by Online Libraries", then "Chemical Sciences", and finally choose Advances in Biochemical Engineering/Biotechnology.

You will find information about the

- Editorial Board
- Aims and Scope
- Instructions for Authors
- Sample Contribution

at springer.com using the search function.

Attention all Users
of the "Springer Handbook of Enzymes"

Information on this handbook can be found on the internet at springeronline.com

A complete list of all enzyme entries either as an alphabetical Name Index or as the EC-Number Index is available at the above mentioned URL. You can download and print them free of charge.

A complete list of all synonyms (more than 25,000 entries) used for the enzymes is available in print form (ISBN 3-540-41830-X).

Save 15%

We recommend a standing order for the series to ensure you automatically receive all volumes and all supplements and save 15% on the list price.

Preface

Organisms are defined by their genetic code – DNA. A detailed knowledge of the mechanisms that duplicate, repair and decode the genetic information is fundamental to our understanding of life itself. Inside the nucleus, DNA is associated with proteins to make chromatin, which forms the template for function. The processes that control chromatin function are presently mostly described in terms of individual DNA–protein interactions.

In reality, however, such interactions are parts of a complex molecular interaction networks that are highly dynamic in time and space. Understanding such complex biological systems requires the unravelling of these networks and catching them in quantitative and predictive models, based on quantitative experiments.

How can a systems biology type of approach be applied to the analysis of the regulation of transcriptional activity? A traditional view of gene expression considers how chromatin structure and transcription factors contribute to the regulation of gene expression by activating RNA synthesis. However, it is becoming increasingly clear that genes do not operate in isolation. Rather, they are components in a network of interactions that orchestrate the use of our genetic information. Although we are still far from understanding the behaviour of these networks, we begin to see the contribution of various aspects. Identifying specific protein–DNA interactions provides a snapshot of gene regulation. Given the transient nature of most cell physiological states, the resulting picture highlights parts of the process of how genes are controlled and regulated. The precise characterization of protein–DNA interactions, the characterization of DNA binding proteins respectively, reveals the components involved in the different steps of gene regulation. Those analyses result in a time-dependent resolution of interactions under a variety of physiological conditions. Knowing the time period during which a regulatory protein is bound to DNA provides insights into the underlying biochemical mechanisms.

The challenge we face at present is to understand how the different systems regulating each of these complex activities can contribute to the global regulation of gene expression. At this stage, the pressing question is how to meet this challenge.

Gene promoters bind numerous transcription factors, which are components of regulatory networks. Their synthesis, activation, modification, cellular localization and turnover are controlled by components of the regulatory network.

A change in gene expression can be obtained by several means: a binding site can become non-functional by mutation, by alternative DNA methylation and histone modification, it can acquire a different function by the same mechanism, or can arise de novo. Even more subtle changes (relative affinity) can result in a different expression pattern of orthologous genes in two tissues, cell lines or species. In conclusion, paralogous genes can acquire different functions by changing their expression pattern or transcriptional responsiveness, whereas their coding sequences can remain identical.

In the field of eukaryotic gene expression, chromatin function and epigenetics we are just at the very beginning of thinking in terms of interaction networks. Our knowledge about gene regulation is increasing rapidly. It is timely to combine the parameter about the chromatin structure and modification, with the genetic information, e.g. methylation, DNA mutations with protein–DNA interaction data, protein network information and affinity data of those interactions.

This book focuses on the fascinating possibilities that emerged during the last years to study protein–DNA interactions in vitro. As an essential part of transcriptional activity during development and in tumorgenesis a separate chapter describing the analysis of DNA modification especially methylation of cytosines at their carbon-5 position was included. Besides the established techniques to characterize protein–DNA interactions by surface plasmon resonance, footprinting techniques and electro-mobility shift assays are still widely used methods. The organism's proteome is a dynamic entity containing thousands of elements involved in numerous intricate networking processes that are related to the developmental stages of the life cycle and to virtually every process in the living cell. A number of methods have been developed to study protein binding to nearly all possible target regions within a genome. To give an insight into those techniques a microarray based method is included. Additionally, a chapter describing in detail the potential of mass spectrometry is included.

The volume editor would like to thank the staff at Springer for help during the preparation of this volume. Special thanks go to Prof. Dr. Thomas Scheper, Institut für Technische Chemie, Hanover University, Germany as well as to Dr. Marion Hertel, chemistry editor, and especially to Ms. Ulrike Kreusel, chemistry desk editor.

Berlin, October 2006   Harald Seitz

Contents

Monitoring Methylation Changes in Cancer
V. Beier · C. Mund · J. D. Hoheisel 1

Analysis of Protein–DNA Interactions
Using Surface Plasmon Resonance
J. Majka · C. Speck 13

Identification of Regulatory Elements by Gene Family Footprinting
and In Vivo Analysis
D. F. Fischer · C. Backendorf 37

Protein Binding Microarrays for the Characterization
of DNA–Protein Interactions
M. L. Bulyk 65

Accuracy and Reproducibility of Protein–DNA Microarray Technology
S. Field · I. Udalova · J. Ragoussis 87

Identification and Characterization of DNA-Binding Proteins
by Mass Spectrometry
E. Nordhoff · H. Lehrach 111

Author Index Volumes 101–104 197

Subject Index 201

Adv Biochem Engin/Biotechnol (2007) 104: 1–11
DOI 10.1007/10_024
© Springer-Verlag Berlin Heidelberg 2006
Published online: 27 October 2006

Monitoring Methylation Changes in Cancer

Verena Beier[1] (✉) · Cora Mund[2] · Jörg D. Hoheisel[1]

[1]Division of Functional Genome Analysis, Deutsches Krebsforschungszentrum, Im Neuenheimer Feld 580, 69120 Heidelberg, Germany
v.beier@dkfz.de

[2]Division of Epigenetics, Deutsches Krebsforschungszentrum, Im Neuenheimer Feld 580, 69120 Heidelberg, Germany

1	Introduction	2
2	Analysis of Genomic DNA Methylation Patterns	2
3	Studying Methylation Patterns by Using Microarrays	4
3.1	Hybridization-based Assay	5
3.2	Primer Extension	8
4	Conclusion and Outlook	10
	References	10

Abstract Methylation of cytosines at their carbon-5 position plays an important role both during development and in tumorgenesis. The methylation occurs almost exclusively in CpG dinucleotides. While the bulk of human genomic DNA is depleted in CpG sites, there are CpG-rich stretches, so-called CpG islands, which are located in promoter regions of more than 70% of all known human genes. In normal cells, CpG islands are unmethylated, reflecting an transcriptionally active state of the respective gene. Epigenetic silencing of tumor suppressor genes by hypermethylation of CpG islands is a very early and stable characteristic of tumorigenesis. The detection of DNA methylation is based on a treatment of genomic DNA with sodium bisulfite, which converts only unmethylated cytosines to uracil, while methylated cytosines stay unaltered. This sequence conversion can be detected in the same way as a single nucleotide polymorphism. Even though different approaches have been established for analysing DNA methylation, so far detection methods that are capable of surveying the methylation status of multiple gene promoters have been restricted to a limited number of cytosines. The use of oligonucleotide microarrays permits the parallel analysis of the methylation status of individual cytosines on a genome-wide and gene-specific level. On the one hand, a hybridization-based setup is described employing microarrays that contain oligonucleotide probes of 17–25 bases in length reflecting the methylated as well as the unmethylated status of each CpG site. After hybridization of sodium bisulfite treated and fluorescently labeled targets, methylation status of individual CpG dinucleotides can be computed based on resulting signal intensities. Secondly, a microarray-based approach for detecting methylation-specific sequence polymorphisms via an on-chip enzymatic primer extension is described.

Keywords CpG island · DNA methylation · Epigenetics · Oligonucleotide microarray · Primer extension

1
Introduction

Epigenetic studies deal with heritable changes of gene expression that are not based on modifications of the DNA sequence like mutation or deletion [1]. Besides chromatin alterations like histone modifications [2], one important epigenetic phenomenon is the methylation of genomic DNA. Formation of DNA methylation patterns is associated with imprinting [3], embryonic development [4] as well as a broad range of human diseases [5]. In the human genome, DNA methylation occurs almost exclusively at cytosine residues, which becomes methylated to 5-methylcytosine in the dinucleotide CpG [6]. The methylation is introduced during replication by a DNA methyltransferase enzyme and is inherited by the daughter cell after a mitotic or meiotic division. In mammalian somatic cells, 5-methylcytosine accounts for about 1% of all bases, varying slightly in different tissue types [7]. Due to an inherent mutability of the methylated cytosine, CpG dinucleotides occur at a much lower level as would be expected based on the GC content of the human genome. Additionally, CpG sites are not evenly spread across the genome. Besides individual CpG dinucleotides all over the genome, there are stretches of several hundred bases in length, which show a high frequency of CpG sites and are referred to as CpG islands. In more than 70% of all known human genes, CpG islands are located in the promoter region and/or within the first exon [8].

Changes in the methylation pattern of genomic DNA belong to the earliest and most consistent features during cancerogenesis [9, 10]. Besides global hypomethylation, which is supposed to be related to chromosomal instability, extensive hypermethylation of CpG islands in promoter regions is observed, which frequently leads to the silencing of genes, including tumor suppressor genes [11, 12], at the transcriptional level. In the last few years, a growing list of genes has been compiled that are methylated in different types of cancer. While genetic mutations are inherited passively by DNA replication, epigenetic modifications are invertible and need to be actively retained. Due to its reversibility, DNA methylation is a promising target for cancer therapeutics. The utilization of DNA methyltransferase inhibitors like 5-aza-2'-deoxycytidine (i.e., decitabine) has been shown to reduce DNA methylation as well as tumor growth [13].

2
Analysis of Genomic DNA Methylation Patterns

The study of methylation of cytosines at a genomic level is traditionally carried out by using methylation-sensitive restriction endonucleases [14]. After digestion, the methylation status can be identified by various methods like PCR, Southern blot analysis or hybridization to a CpG island microarray [15]. An alternative strategy is the complete digestion of genomic DNA with a subsequent determination of the proportion of methylated cytosines by high-performance liquid chromatography (HPLC) [16], capillary electrophoresis [17] or mass spectrometry [18, 19]. While the former approach is limited to specific restriction sites, the latter results only in information on the all-over methylation level of the genome, without focusing on the methylation of single genes. Moreover, both methods require large amounts of genomic DNA.

For addressing the detection of changes in the DNA methylation pattern of individual genes, methods have been established that are based on the bisulfite treatment of genomic DNA. Sodium bisulfite deaminates unmethylated cytosine to uracil and, upon PCR amplification, to thymine. In contrast, 5'-methylcytosine stays unaltered and thus becomes cytosine upon subsequent PCR amplification. In consequence, methylation patterns can be detected in principle in the same way as single nucleotide polymorphisms (SNPs). Among different methods for detecting methylation changes, methylation-specific PCR (MSP) [20] is the most prominent assay due to fact that the experimental set-up of this method is easy to realisable and does not make great demands on technical equipment. By using two sets of primers – one directed against the methylated, C-containing template, the other against the unmethylated, T-containing sequence – the methylation status of the DNA is determined. Slightly modified approaches are based on fluorescent real-time PCR (MethyLight) [21], a method which can be improved by adding methylation-specific blockers (HeavyMethyl) [22]. An alternative assay is the combined bisulfite restriction analysis (COBRA) [23], which uses PCR amplification after sodium bisulfite treatment, followed by restriction digestion and quantification of the resulting fragments. Conservation of pre-existing or creation of new restriction sites displays alteration in methylation. Another technique called methylation-sensitive single-nucleotide primer extension (Ms-SNuPE) [24] also uses a PCR amplicon of sodium bisulfite converted genomic DNA as template. Oligonucleotide probes anneal to the sequence 5' upstream to the CpG site being monitored. A single-nucleotide primer extension takes place in the presence of a DNA polymerase and an appropriate radioactively labeled dNTP. Gel electrophoresis is used for visualization of the reaction products, followed by phosporimage analysis. All these PCR-based methods access only a limited number of CpG sites and can be applied to only one gene per experiment. Bisulfite sequencing [25, 26] on the other hand, which implies either direct sequencing or the sequencing of several subclones of an amplicon of bisulfite treated genomic DNA, permits the analysis of every CpG dinucleotide on defined DNA fragments of several hundred nucleotides in length and thus offers a high degree of resolution. An alternative strategy for the quantitative analysis of methylation in multiple CpG sites is Pyrosequencing® [27–29]. Starting from single stranded, bisulfite converted DNA templates, complementary strands are synthesized by adding sequentially the four nucleotides. Every successful nucleotide incorporation leads to the release of pyrophosphate, whose following enzymatical conversion conducts light emissions in a quantity that is proportional to the number of incorporations. However, the experimental procedure of both sequencing approaches is again limited to single genes per assay and is both cost-intensive and time-consuming. Recently, a highly parallel sequencing system using a pyrosequencing protocol in high-density picolitre-sized reactors was established [30]. The system performs sequencing by synthesis of hundreds of thousands fragments of 80–120 bases in length simultaneously in open wells of a fibre-optic slide and therefore reaches an approximately 100-fold increase in throughput over current Sanger sequencing technology. Still being connected with with a relatively high effort in order to get quantitative data of genomic DNA methylation, this method represents an efficient tool for bisulfite sequencing.

3
Studying Methylation Patterns by Using Microarrays

By use of oligonucleotide microrarrays, the parallel analysis of numerous CpG dinucleotides on a gene-specific as well as genome-wide level becomes possible.

Recently, several approaches using spotted oligonucleotide-based microarrays have been published. On one hand, target DNA is generated by differential cleavage with methylation-sensitive restriction endonucleases followed by a ligation-mediated PCR amplification [15]. DNA derived from tumor tissue and from normal tissue, respectively, is labeled with two different fluorescent dyes and hybridized simultaneously to a microarray containing genomic DNA fragments that predominantly cover CpG islands, resulting in a direct comparison of the methylation situation. Resolution of this method is about 0.5–3 kilobases and is limited to loci with pairs of probed restriction sites. In contrast, a high resolution method without limitation to certain sequences is the hybridization of bisulfite-converted target DNA against probes of various length between 17 and 23 nucleotides that are designed to match either with methylated or unmethylated DNA targets [31, 32]. Differences in hybridization efficiency of full-match versus mismatch binding are used to discriminate between methylated and unmethylated status. Results revealed the capability of microarrays to detect methyation profiles which can be used for tumor classification [31]. However, the selection of the oligonucleotide probes is a critical step, because these probes should be highly specific to the respective polymorphism and furthermore should show same the hybridization behaviour. Due to these limitations, the complexity of the described arrays is rather low and still restricted to individual CpG dinucleotides in single genes. Recently, a method for profiling genomic DNA methylation using universal bead arrays was described [33]. Bisulfite treated, biotinylated genomic DNA is immobilized on paramagnetic beads. For every CpG site, four query oligonucleotide probes are designed. Two of them have a 3′ portion specific to either the methylated or the unmethylated status of the CpG site. The 5′ end consists of an universal PCR primer sequences for the methylated status and the unmethylated status, respectively (P1 and P2). Two other oligonucleotide probes have a common universal PCR primer sequence at their 3′ end (P3), an address sequence in the middle, which is complementary to a capture sequence on the array, and a CpG locus specific 5′ sequence, which hybridizes several bases downstream to the interrogated CpG site and therefore in proximity to the first two oligonucleotide probes. After hybridization of the pooled query oligonucleotides, methylation specific extension followed by ligation of the query oligonucleotide probes takes place, resulting in amplifiable templates. Amplification is performed by using the common primers P1, P2 and P3. Since P1 and P2 are labeled with two different fluorescent dyes, a discrimination of methylated and unmethylated targets after hybridization of the amplicon to a microarray becomes possible. This multiplexed method enables the quantitative measurement of DNA methylation at up to 1536 different CpG sites simultaneously. Since array results are read out on a universal bead array, a high flexibility of the genes or CpG sites to be studied is warranted.

3.1
Hybridization-based Assay

Recently, we established a hybridization-based assay on in situ synthesized oligonucleotide microarrays of febit [34]. The light-directed, mask-free oligonucleotide synthesis of this flexible platform enables the generation of microarrays containing probes for every CpG dinucleotide within any genomic region of interest. Hence, the study of the methylation status of multiple CpG sites in a highly parallel fashion is feasible [35] (Fig. 1). With a microarray-on-demand system in the user's lab, adding new genes or CpG sites can be done directly within the next experiment and does not require longer planning or ordering of new pre-synthesized capture probes. Even though array design can be optimized empirically, there might be CpG sites which are difficult to get due to sequence dependent differences in dissociation behaviour and specificity of the oligonucleotide probes, a fact that is true for all assays based on the formation of a stable DNA duplex between target DNA and capture probes. The performance of melting curve analysis enabled with febit's Geniom® technology might help to overcome the problem of finding specific probe sequences being able to examine every CpG dinucleotide.

The design of the microarrays consists of short oligonucleotide probes of 17–25 nucleotides in length. Every CpG dinucleotide is represented by two sequence variations of the oligonucleotide probes: one oligomer querying

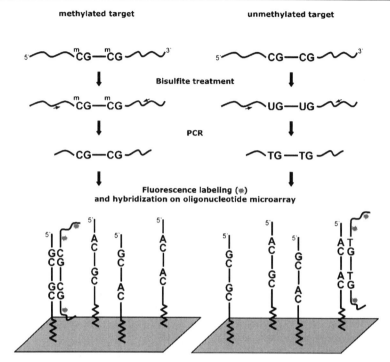

Fig. 1 Schematic outline of the analysis of genomic DNA methylation by use of oligonucleotide microarrays. Genomic DNA is treated with sodium bisulfite and the region of interest is then amplified by PCR. The amplicon is labeled fluorescently and hybridized to the microarrays containing oligomer probes interrogating the different methylation states of every single CpG site. *Left*, hybridization of a completely methylated target is shown, forming a perfect match with the corresponding probe containing CG only. *Right*, the unmethylated, TG-containing target binds to the complementary AC-containing oligomer probe

the unmethylated status (U) contains the sequence TG, while for the methylated status (M) a CG-motif is used instead. If oligomer probes cover more than one CpG site, probes for all possible combinations of the methylated and the unmethylated status of the adjacent CpG sites are generated, therefore resembling a set of 2^n different probe sequences for a number of n CpG dinucleotides within the oligomer probe sequence. In contrast, Bibikova and colleagues as well as almost every other scientists using microarray platforms make the assumption that neighboring CpG sites have the same methylation status as the site of interest, which leads to a drastically reduced sensitivity in all cases which do not follow this assumption. For making the enquiry of every single CpG dinucleotide possible, probe design is further done in a way that each CpG is located once in the central position of the respective probe set, allowing for the best discrimination power between methylated and unmethylated status of this CpG site. To enable the analysis of both strands of a double-stranded target, for each sense probe the reverse complementary antisense probe is generated, too.

For hybridization, genomic DNA is isolated from the tissue to be studied and treated with sodium bisulfite. Subsequently, regions of interest are amplified by PCR. For ensuring an unbiased amplification of unmethylated and methylated regions, primers without CpG sequence motifs are selected. The amplicons are labeled with a fluorescent dye and hybridized to the microarray.

Based on the resulting hybridization signals (Fig. 2), the individual methylation level of every CpG dinucleotide is calculated from the intensity ratio M/M + U of the respective methylated (M) and unmethylated (U) probes. As expected, performance of the individual oligomer probes exhibit a wide diversity. While many probes are able to discriminate very well between different methylation levels, others clearly fail. This performance difference can be attributed to reasons like cross-hybridization or a low hybridization efficiency. For selection of oligonucleotide probes with a high discriminatory power, a calibration is performed for each probe on the basis of hybridizations with reference samples. As a positive control for fully methylated DNA, genomic DNA treated with SssI methylase is used. SssI methylase quantitatively methylates all cytosines of CpG sites to 5-methylcytosine, which stay therefore unaltered in the subsequent sodium bisulfite treatment. As a negative control for unmethylated DNA, regions of interest are amplified prior to bisulfite treatment, resulting in a loss of methylation. Treatment of these PCR amplicons with sodium bisulfite leads to a full conversion of all CG sequences to TG, thus reflecting completely unmethylated DNA. By hybridization of defined mixtures of the positive and the negative controls, simulating

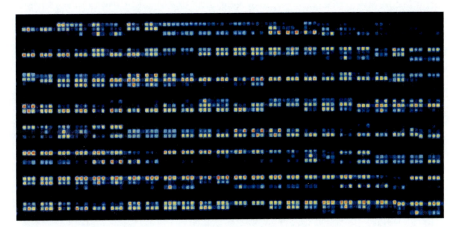

Fig. 2 Typical image of a part of an oligonucleotide microarray for epigenetic analysis based on hybridization. Signal intensities are acquired and taken for calculation of the methylation levels

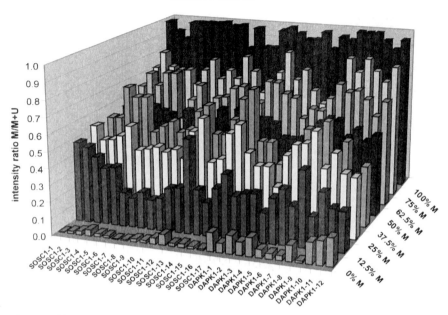

Fig. 3 Calibration data of 17 oligonucleotide probes selected for the gene SOCS1 and 12 probes for the gene DAPK1. For every oligomer probe, a calibration curve is recorded. The diagram shows the intensity ratio and thus the methylation levels of DNA mixtures that reflect 0%, 12.5%, 25%, 37.5%, 50%, 62.5%, 75% and 100% methylation. Plotted are only oligonucleotide probes, which clearly discriminate between the different methylation levels

methylation levels of 0%, 25%, 50%, 75%, and 100%, calibration curves for all oligonucleotide probes of an initial design are recorded, and thus the linearity of the system is evaluated (Fig. 3). Only those oligonucleotide probes, which fulfill the criterion of being able to efficiently distinguish between different methylation levels are selected for further analysis of patient samples.

3.2
Primer Extension

Besides the hybridization-based assay, we are using on-chip enzymatic reactions for detecting methylation-specific single nucleotide polymorphisms. The strategy of using a polymerase-based primer extension reaction benefits from the fact that combining the specificity of hybridization and enzymatic reaction for distinguishing between different methylation states is better by a factor of 10- to 100-fold than hybridization-based discrimination [36]. While the oligonucleotide probes for conventional hybridization-based microarrays are synthesized in 3′–5′ direction, the synthesis direction has to be inverted for synthesizing probes for an enzymatic elongation accordingly to the direction of biological synthesis, resulting in oligomers with a freely accessible 3'-OH-end [37]. Oligonucleotide probes of 25 bases in length are designed to match the flanking sequence of the CpG site on both strands, sense and antisense. Extension with a fluorescently labeled dideoxynucleotide can only occur, if the base added is complementary to the annealed target sequence at the position of the CpG site. In case of a methylated, CG-containing target, elongation is performed with ddGTP, while for an unmethylated and thus TG-containing target, ddATP is used (Fig. 4).

Using double-stranded PCR amplicons as target, oligomer probes querying the sense as well as the antisense strand are on the microarray. Having only two possible reactions in either strand, study of cytosine methylation is less complex than traditional genotyping with potential incorporation of four different bases. Compared to our hybridization-based epigenetic arrays,

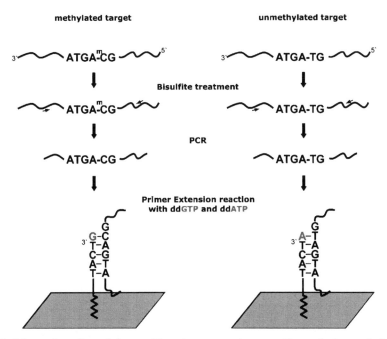

Fig. 4 Schematic outline of the on-chip primer extension assay for analysing methylation patterns. Oligonucleotide probes are designed to match the target sequence 3' downstream to the polymorphism site (only one target strand is depicted). Bisulfite treatment of genomic DNA is followed by PCR amplification of the genes of interest. Adding target DNA together with polymerase reaction mixture onto the microarray, elongation occurs specifically with fluorescently labeled dideoxynucleotides complementary to the annealed sequence. *Left*, annealing of the methylated CG containing sequence leads to an elongation with ddGTP. *Right*, the complementary TG containing unmethylated sequence causes incorporation of ddATP

the number of oligonucleotide probes required for the analysis of every single CpG dinucleotide within a given sequence is reduced, since discrimination of the methylation-dependent sequence variation is based only on the sensitivity of the polymerase reaction regardless of a less stable duplex formation due to mismatches in adjacent CpG sites.

4 Conclusion and Outlook

By characterisation of genome wide and gene-specific DNA methylation patterns fundamental insights into tumor development will be obtained. This especially holds true when the epigenetic data are compared with available clinical data as well as results from corresponding gene expression profiling studies. By establishing a system that can be used for prognosis as well as diagnosis, a basis of a profound epigenetic tumor classification is provided. Besides using epigenetic arrays as a diagnostic tool, monitoring therapeutical interventions before, during and after cancer therapy becomes feasible.

Acknowledgements We are grateful to Markus Beier and Frank Lyko for helpful discussions. Funding by the Deutsche Forschungsgemeinschaft (DFG) and the NGFN programme of the German Federal Ministry of Education and Research (BMBF) is gratefully acknowledged.

References

1. Wu CT, Morris JR (2001) Science 293:1103
2. Peterson CL, Laniel MA (2004) Curr Biol 4:R546
3. Paulsen M, Ferguson-Smith AC (2001) J Pathol 195:97
4. Monk M, Boubelik M, Lehnert S (1987) Development 99:371
5. Rodenhiser D, Mann M (2006) CMAJ 174:341
6. Bird A (2002) Genes Dev 16:6
7. Ehrlich M et al. (1982) Nucleic Acids Res 10:2709
8. Saxonov S, Berg P, Brutlag DL (2006) Proc Natl Acad Sci USA 103:1412
9. Jones PA, Baylin SB (2002) Nat Rev Genet 3:415
10. Warnecke PM, Bestor TH (2000) Curr Opin Oncol 12:68
11. Baylin SB, Hermann JG (2000) Trends Genet 16:168
12. Esteller M, Sanchez-Cespedes M, Rosell R, Sidransky D, Baylin SB, Hermann JG (1999) Cancer Res 59:67
13. Lyko F, Brown R (2005) J Natl Cancer Inst 97:1498
14. Cedar H, Solage G, Glaser G, Razin A (1979) Nucleic Acids Res 22:2125
15. Yan PS, Chen C-M, Shi H, Rahmatpanah F, Wei SH, Caldwell CH, Huang TH (2001) Cancer Res 61:8375
16. Eick D, Fritz HJ, Doerfler W (1983) Anal Biochem 135:165
17. Stach D, Schmitz OJ, Stilgenbauer S, Benner A, Döhner H, Wiessler M, Lyko F (2003) Nucleic Acids Res 31:e2

18. Babinger P, Kobl I, Mages W, Schmitt R (2001) Nucleic Acids Res 29:1261
19. Friso S, Choi SW, Dolnikowski GG, Selhub J (2002) Anal Chem 74:4526
20. Hermann JG, Graff JR, Myohanen S, Nelkin BD, Baylin SB (1996) Proc Natl Acad Sci USA 93:9821
21. Eads CA, Danenberg KD, Kawakami K, Saltz LB, Blake C, Shibata D, Danenberg PV, Laird PW (2000) Nucleic Acids Res 28:e32
22. Cottrell SE, Distler J, Goodman NS, Mooney SH, Kluth A, Olek A, Schwope I, Tetzner R, Ziebarth H, Berlin K (2004) Nucleic Acids Res 32:e10
23. Xiong Z, Laird PW (1997) Nucleic Acids Res 25:2532
24. Gonzalgo ML, Jones PA (2002) Methods 27:128
25. Frommer M, McDonald LE, Millar DS, Collis CM, Watt F, Grigg GW, Molloy PL, Paul CL (1992) Proc Natl Acad Sci USA 89:1827
26. Clark SJ, Harrison J, Paul CL, Frommer M (1994) Nucleic Acids Res 22:2990
27. Uhlmann K, Brinckmann A, Toliat MR, Ritter H, Nürnberg P (2002) Electrophoresis 23:4072
28. Colella S, Shen L, Baggerly KA, Issa J-PJ, Krahe R (2003) Biotechniques 35:146
29. Tost J, Dunker J, Gut IG (2003) Biotechniques 35:152
30. Margulies M, Egholm M, Altman WE, Attiya S, Bader JS, Bemben LA, Berka J, Braverman MS, Chen YJ, Chen Z, Dewell SB, Du L, Fierro JM, Gomes XV, Godwin BC, He W, Helgesen S, Ho CH, Irzyk GP, Jando SC, Alenquer ML, Jarvie TP, Jirage KB, Kim JB, Knight JR, Lanza JR, Leamon JH, Lefkowitz SM, Lei M, Li J, Lohman KL, Lu H, Makhijani VB, McDade KE, McKenna MP, Myers EW, Nickerson E, Nobile JR, Plant R, Puc BP, Ronan MT, Roth GT, Sarkis GJ, Simons JF, Simpson JW, Srinivasan M, Tartaro KR, Tomasz A, Vogt KA, Volkmer GA, Wang SH, Wang Y, Weiner MP, Yu P, Begley RF, Rothberg JM (2005) Nature 437:376
31. Adorjan P, Distler J, Lipscher E, Model F, Muller J, Pelet C, Braun A, Florl AR, Gutig D, Grabs G, Howe A, Kursar M, Lesche R, Leu E, Lewin A, Maier S, Muller V, Otto T, Scholz C, Schulz WA, Seifert HH, Schwope I, Ziebarth H, Berlin K, Piepenbrock C, Olek A (2002) Nucleic Acids Res 30:e21
32. Gitan RS, Shi H, Chen CM, Yan PS, Huang TH (2002) Genome Res 12:158
33. Bibikova M, Lin Z, Zhou L, Chudin E, Garcia EW, Wu B, Doucet D, Thomas NJ, Wang Y, Vollmer E, Goldmann T, Seifart C, Jiang W, Barker DL, Chee MS, Floros J, Fan JB (2006) Genome Res 16:383
34. Baum M, Bielau S, Rittner N, Schmid K, Eggelbusch K, Dahms M, Schlauersbach A, Tahedl H, Beier M, Guimil R, Scheffler M, Hermann C, Funk JM, Wixmerten A, Rebscher H, Honig M, Andreae C, Buchner D, Moschel E, Glathe A, Jager E, Thom M, Greil A, Bestvater F, Obermeier F, Burgmaier J, Thome K, Weichert S, Hein S, Binnewies T, Foitzik V, Muller M, Stahler CF, Stahler PF (2003) Nucleic Acids Res 31:e151
35. Mund C, Beier V, Bewerunge P, Dahms M, Lyko F, Hoheisel JD (2005) Nucleic Acids Res 33:e73
36. Pastinen T, Kurg A, Metspalu A, Peltonen L, Syvänen AC (1997) Genome Res 7:606
37. Beier M, Stephan A, Hoheisel JD (2001) Hel Chim Acta 84:2089

Analysis of Protein–DNA Interactions Using Surface Plasmon Resonance

Jerzy Majka[1] · Christian Speck[2] (✉)

[1] Department of Biochemistry and Molecular Biophysics,
Washington University School of Medicine, St. Louis, MO 63110, USA

[2] Cold Spring Harbor Laboratory, 1 Bungtown Road, Cold Spring Harbor, NY 11724, USA
Speck@cshl.edu

1	Introduction	14
2	Surface Plasmon Resonance	15
2.1	General Principle	15
2.2	BIAcore	16
3	Guidelines for Studying Protein–DNA Interactions with the BIAcore . . .	18
3.1	Knowledge of the Experimental System	18
3.2	Ligand Considerations	18
3.3	Mass Transport Limitations	19
3.4	Chip Surface Considerations	20
3.5	The Reference Surface and Control Injections	20
3.6	Regeneration of the Chip Surface	20
3.7	Buffer Conditions	21
3.8	The Experiment	21
4	Analysis of Data	21
4.1	1 : 1 Interactions Analyzed by Quantitative Analysis	22
4.1.1	Introduction	22
4.1.2	Association	23
4.1.3	Dissociation	24
4.1.4	Equilibrium	25
4.1.5	Practical Guideline on Evaluation of the Rate Constants k_{on} and k_{off} and the Dissociation Constant K_d	25
4.2	Kinetic Analysis of Interactions Between Protein and a DNA Containing Multiple Protein Binding Sites	27
4.2.1	Introduction	27
4.2.2	Curve Fitting with the BIAevaluation Software	27
4.2.3	Hill's Cooperativity Factor	28
4.2.4	Cooperativity Fold α Factor	30
5	Qualitative Analysis	31
5.1	Stoichiometry Determination	31
5.2	Qualitative Analysis of Protein–DNA Complex Formation	32

6	Surface Plasmon Resonance to Study Protein–DNA Interactions 1993–2006	32
7	Technological Developments	35
References		36

Abstract Protein–DNA interactions are required for access and protection of the genetic information within the cell. Historically these interactions have been studied using genetic, biochemical, and structural methods resulting in qualitative or semiquantitative interaction data. In the future the focus will be on high quality quantitative data to model a huge number of interactions forming a specific network in system biology approaches.

Toward this aim, BIAcore introduced in 1990 the first commercial machine that uses surface plasmon resonance (SPR) to study the real-time kinetics of biomolecular interactions. Since then systems have been developed to allow for robust analysis of a multitude of protein–DNA interactions. Here we provide a detailed guide for protein–DNA interaction analysis using the BIAcore, starting with a description of the SPR technology, giving recommendations on preliminary studies, and finishing with extensive information on quantitative and qualitative data analysis. One focus is on cooperative protein–DNA interactions, where proteins interact with each other to modulate their binding specificity or affinity. The BIAcore has been used for the last 14 years to study protein–DNA interactions; our literature review focuses on some high quality studies describing a wide range of experimental uses, covering simple 1 : 1 interactions, analysis of complicated multiprotein–DNA interaction systems, and analytical uses.

Keywords SPR · Kinetic · Cooperative · BIAcore · DnaA

1
Introduction

With the discovery of the DNA structure [1] more than 50 years ago we started to understand that the genetic information of the cell is stored in DNA. By now this genetic information is available for the human species and many other species due to large-scale DNA sequencing projects. The cell accesses genetic information stored in DNA through specialized proteins, which can bind to DNA. Frequently, numerous different proteins collaborate to build sophisticated machines on top of DNA, which are required for replication, transcription, modification, and repair of DNA. To understand the complex interaction between proteins, protein complexes, and DNA ultimately a map of protein–DNA and protein–protein interactions has to be developed. Systems biology is working toward this goal, by defining in quantitative terms the nature of interactions and by incorporating this information with the help of bioinformatics into a model.

Several methods have been developed to study protein–DNA interactions with the focus being to define the interaction in qualitative or quantitative terms. Historical methods that measure protein–DNA interactions like footprinting [2] and gel shifts [3] are qualitative or semiquantitative, because they are not compatible with the fast association rates many DNA binding proteins display. This led to the development of biophysical methods, which rely on fast optical systems for quantitative protein–DNA interaction measurements. Two of the most important optical methods are surface plasmon resonance (SPR) and rapid-scanning stopped-flow spectrophotometry. SPR is used to monitor the interaction between biomolecules. One molecule is immobilized on a chip surface and binding of the other molecule is detected as a change of refractive index on the chip surface. The interaction can be monitored very accurately in real time. Since the change in refractive index corresponds to a change in mass this method can also yield data on the stoichiometry of complexes in addition to binding kinetics. Stopped-flow spectrophotometry measures a change in absorbance as a function of reaction time. It has the advantage that the molecules are freely moving in solution. Not every biomolecule will show a change in absorbance, therefore one molecule frequently has to be labeled with a fluorochrome, which can interfere with the activity of the molecule. Both methods can monitor fast interactions in real time and are able to deliver the high quality quantitative data required for systems biology. However, due to the ease of use, low requirements in respect of the biomolecule, and simultaneous analysis of multiple interaction partners, SPR is the most successful. Consequently this review will focus on SPR as a tool to study protein–DNA interactions.

2
Surface Plasmon Resonance

2.1
General Principle

SPR is a physical phenomenon which occurs when a polarized light beam is projected through a prism onto a thin metal film (gold or silver) (Fig. 1). At a specific angle of the projected light, resonance coupling between light photons and surface plasmons of the gold can occur since their frequencies match. Because the resonance leads to an energy transfer, the reflected light shows a sharp intensity drop at the angle where SPR is taking place. Resonance coupling of the plasmons generates an evanescent wave that extends 100 nm above and below the gold surface. For SPR as an analytical tool it is most important that a change in the refractive index within the environment of the evanescent wave causes a change of the angle where the sharp intensity drop can be observed (Fig. 1, compare angle 1 and angle 2). Binding of one biomolecule to another immobilized on top of the sensor chip's gold surface will lead to a change of refractive index and will be recorded as a change

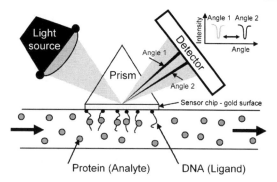

Fig. 1 The basic principles of surface plasmon resonance and analysis in BIAcore experiments. See text for details

in the reflected light by a detector. This setup enables real-time measurement of biomolecular interactions, with refractive index changes proportional to mass changes.

2.2
BIAcore

BIAcore AB is the main supplier of automated SPR detection systems (BIAcore 3000, T100, and more, see www.BIAcore.com); other suppliers are Texas Instruments (Spreeta and TISPR-I) and Nippon Laser and Electronics (SPR-670). In 2004, 88% of publications that reported the use of SPR indicated that the data were generated with BIAcore instruments [4]. Therefore, in the following the focus will be on BIAcore instruments and BIAcore nomenclature will be used.

In our case DNA (ligand) is immobilized on the chip surface (Fig. 2). A constant buffer flow over the chip surface precedes injection of a protein (analyte). Binding of the protein to DNA can be observed as a change in the position of the reflected light minimum, which can be translated into a change of resonance units (RUs); 1 RU corresponds to a change in angle of 0.0001° and 1 pg mm^{-2} of protein. The RUs will be recorded and displayed in real time by a computer. Once the reaction reaches an equilibrium, the binding and dissociation rates of the protein on DNA will be equal. Subsequent to the protein injection, buffer flow will follow and dissociation of the protein will be monitored. Therefore, the observed minimum in the reflected light will gradually return to its original position, before protein injection. At any point of the dissociation phase a regeneration fluid can be injected into the system, which will allow rapid release of any protein from the chip surface prior to a new round of analyte injection. Accordingly, a real-time interaction experiment will contain the following parts: an association

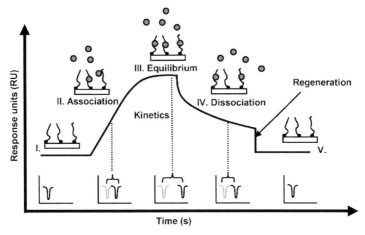

Fig. 2 The sensorgram is a graphical representation of the signal produced during a binding experiment. The sensor chip carries immobilized DNA. I. Initially buffer is running over the sensor chip, and no change in signal is observed. II. A DNA binding protein is injected and associates with the DNA. This leads to a change in the refractive index, which is monitored in the BIAcore as a change in angle where the drop in intensity of the reflected light is observed. This change in angle is visualized in the sensorgram. III. Association and dissociation of the protein and the DNA occur at equal rates during the equilibrium. IV. Protein injection is replaced by buffer flow and dissociation of the protein from the DNA is observed. This leads to a reduction in refractive index and the angle where the drop of intensity is observed. V. At any point of the dissociation regeneration liquid can be injected, which will remove all the protein from the chip surface

phase, with the protein binding to the immobilized DNA; an equilibrium phase, with equal association/dissociation rates for the protein; and a dissociation phase, with buffer flow washing away any protein that dissociates from the DNA.

The standard BIAcore instrument contains a sensor chip with four flow cells, which enables measurement of four different ligands at a time. One side of the chip is attached to the SPR detection system as discussed. On the other side is a microfluidic system, which forms four parallel channels on the sensor chip surface and facilitates buffer flow and injections. The standard sensor chip carries on top of the gold layer a carboymethylated dextran surface, which is used for immobilization of biomolecules. This surface is characterized by low nonspecific binding, except for very basic compounds, which demands consideration since DNA binding proteins frequently carry a basic charge. Several other sensor chips are available, including chips with immobilized streptavidin for easy capture of biotinylated DNA fragments and chips with reduced negative charge, which can help to analyze extremely basic proteins.

3
Guidelines for Studying Protein–DNA Interactions with the BIAcore

The BIAcore technology can be used to study binding constants, stoichiometry, and thermodynamics (discussed in detail in [5]) of protein–DNA interactions; however, experimental design has to be considered to produce meaningful results.

3.1
Knowledge of the Experimental System

To study a novel protein–DNA interaction with the BIAcore an initial characterization of the interaction by a gel-shift and/or footprint assay will be required [2, 3]. These assays provide data on the stoichiometry of the protein–DNA complex, the specificity of the interaction, and the localization of the binding site on the DNA.

Frequently, proteins bind to DNA following a 1 : 1 interaction model, which can be observed in a gel retardation assay as a single shifted band. The protein concentration required to produce a gel shift can be used to estimate a range of protein concentrations for BIAcore experiments. As a rule of thumb one should use protein concentrations in the range of tenfold below and above the apparent dissociation constant K_d [3]. A footprint assay can be used to localize the binding site of the protein on the DNA. Consequently, a minimal DNA fragment containing only the binding site can be generated for analysis in the BIAcore. This is important, since additional DNA surrounding the binding site may contain more binding sites or may increase the chance of nonspecific protein–DNA interactions. The footprint assay can also identify whether a protein interacts nonspecifically with the DNA, which needs to be considered in further experimental design.

3.2
Ligand Considerations

The ligand has to be immobilized on a sensor chip surface by either chemical cross-linking or with an affinity interaction like biotin–streptavidin (sensor chip SA), nickel chelate-His-Tag (sensor chip NTA) or antibody–antigen (sensor chip CM5). To ensure the measurement of accurate data the immobilization procedure should not interfere with the ligand–analyte interaction. The most common immobilization procedure for proteins is chemical cross-linking, which can reduce the activity or affinity of the interaction due to the modification of the protein or by blocking the binding site.

Immobilization of DNA, on the other hand, can be achieved with biotinylated DNA and streptavidin sensor chips, which results in accessible DNA and a uniform attachment of the ligand. Due to the strong interaction of biotin and streptavidin these chips withstand fairly harsh regeneration conditions, e.g., high salt and low concentrations of sodium dodecyl sulfate (SDS). The size of the immobilized DNA should correspond to the length of the DNA footprint plus a few extra bases (3–6 bp) on each side as a spacer. The biotin should be positioned at the very end of the DNA so it will not interfere with binding. The DNA can be generated by the polymerase chain reaction (PCR) or through annealing of two complementary oligonucleotides. PCR fragments require gel purification of nonincorporated biotinylated primer and smaller PCR-failure fragments. If the double-stranded DNA (dsDNA) is produced by annealing of two oligonucleotides, the nonbiotinylated oligonucleotide should be in tenfold molar excess to ensure that all biotinylated DNA oligonucleotide is in the form of dsDNA.

3.3
Mass Transport Limitations

Mass transport limitation occurs when the association of an analyte (protein) to the ligand (DNA) is limited by the diffusion of analyte to the surface of the chip.

Most protein–DNA interactions are characterized by very fast association rates. If a protein has an association rate constant (k_{on}) above 1×10^6 M^{-1}s^{-1} it will likely be limited by mass transport [6]. This can be reduced by several means:

1) The amount of DNA immobilized should be as low as possible. As a rule of thumb the response of the protein should not exceed 100 RU—the range of 30–50 RU still works well.
2) The flow rate should be very high, $\geq 50\ \mu l\ min^{-1}$.
3) New BIAcore systems such as BIAcore 3000 or T100 use a different flow cell geometry than the older systems BIAcore 2000 or BIAcore X, further reducing mass transport effects.
4) No glycerol or sucrose should be present in the buffer or sample since this reduces the diffusion rate.

If, after following all these optimization steps, a kinetic reaction is still restricted by mass transport a mass transport rate constant (k_t) can be incorporated into the binding model.

$$PROTEIN_{sol} \xrightarrow{k_t} PROTEIN_{surf} + DNA \underset{k_{off}}{\overset{k_{on}}{\rightleftarrows}} PROTEIN \cdot DNA\ ,$$

where $PROTEIN_{sol}$ is the protein concentration in the flow cell, $PROTEIN_{surf}$ is the protein concentration at the chip surface accessible for interaction with DNA, and k_t is the mass transport rate constant. This formula has been incorporated into the BIAcore evaluation software as a modified evaluation model calculating, additionally to the kinetic parameters, the mass transport rate constant k_t.

3.4
Chip Surface Considerations

As discussed in Sect. 2.2, the carboxymethylated dextran surface of the standard BIAcore sensor chip CM5 or SA (CM5 with immobilized streptavidin) has a negative charge at neutral pH. This can be problematic with very basic proteins, like many DNA binding proteins, since they will bind nonspecifically to the dextran. By reducing the pH of the buffer the negative charge of the dextran will be reduced; however, at the same time the basic protein might become more positively charged due to the lower pH. Consequently, the nonspecific binding potential should be tested for basic proteins, by scouting the pH of the buffer in the range of 6–8. Increasing the salt concentration will also reduce nonspecific interactions with the matrix [5, 7]. Finally, a different chip with a lower degree of carboxymethylation can be used—sensor chip CM4. To study extremely basic proteins like histones a technique was reported to reverse the charge of the chip surface by absorbing polyethylenimine (PEI) on the surface of sensor chip CM5 [8]. This procedure abrogates the strong nonspecific binding of histones to the sensor chip.

3.5
The Reference Surface and Control Injections

A sensor chip usually contains four flow cells (BIAcore 3000, T100). One flow cell is always used as a reference cell, which can correct for changes from the injection procedure, nonspecific binding, or baseline drift. The reference flow cell should mimic the specific flow cell as much as possible. If a specific protein–DNA complex is studied, another DNA of the same length with unrelated DNA sequence should be used as a reference and the amount of immobilized DNA should be identical for the reference and specific flow cell. However, the use of a reference flow cell cannot always correct for changes in the baseline throughout a long experiment. Therefore buffer-only injections are spaced throughout the experiment and serve as a reference that will be subtracted from the actual data set to obtain high quality data [9].

3.6
Regeneration of the Chip Surface

Frequently, after the protein is injected, a fraction of the protein stays nonspecifically attached to the sensor chip. To remove all protein from the chip a regeneration fluid, which usually contains high salt levels, detergents, or has extreme pH, will be injected. Good regeneration conditions remove the analyte completely from the surface without removing or damaging the immobilized ligand. Myszka [10] has reviewed a wide range of regeneration conditions. For gentle but fast regeneration we use a combination of high salt and a low concentration of the ionic detergent SDS (10 mM Tris-HCl pH 7.5, 500 mM NaCl, 1 mM EDTA, 0.005% SDS).

3.7
Buffer Conditions

Protein–DNA interactions are dominated by ionic interactions, and therefore the buffer system for the interaction analysis needs to be chosen carefully. It has been observed that the ionic strength of the buffer influences the specific interactions less than the nonspecific protein–DNA interactions. Therefore, increasing the salt concentration can decrease the nonspecific binding to an undetectable level, keeping the specific interaction intact [7]. Our standard running buffer for studying protein–DNA interactions contains: 25 mM HEPES pH 7.6, 100 mM K acetate, 1 mM Mg acetate, 0.005% BIAcore Surfactant P20, 10 mg ml^{-1} BSA, poly-dIdC, and poly-dAdT (33 ng µl^{-1}). For gel-shift and footprint assays, competitor DNA is frequently added to reduce nonspecific protein DNA interactions. The same effect has been observed in our own BIAcore experiments, especially at high protein concentrations.

3.8
The Experiment

Running buffer is prepared fresh, filtered, and degassed at the same temperature as that at which the experiment will be conducted—usually 22 °C. The machine is primed three times to equilibrate the system to the new buffer. Samples spanning concentrations tenfold above and below the K_d, estimated from gel shifts or footprints, are diluted on ice in running buffer and kept at 4 °C. Buffer-only injections evenly spaced throughout the experiment are recommended. One after another the samples are injected using the "kinject" [11] command using a high flow rate (50–100 µl min^{-1}). At the end of each cycle, bound protein is removed with a 5-s pulse of regeneration fluid (Sect. 3.6). To equilibrate the system after regeneration back to the running buffer, 150 µl of buffer is passed through the system before the next sample injection.

4
Analysis of Data

Before data analysis can start the response curves generated during the experiment require processing. The unwanted part of a sensorgram, such as a very long baseline before injection and regeneration, can be removed. The baseline must be adjusted to zero since most of the fitting algorithms require that. The response curves must be aligned, usually by setting $t = 0$ s as the start of the injection for each curve, and finally the response in the reference cell should be subtracted (see BIAcore handbook [11] and Myszka [9] for details). The subtraction can be carried out "on line" during the experiment or afterwards. Often, even though both curves were aligned on the time axis, the subtraction of the reference cell data results in the appearance of spikes in the sensorgrams, especially at the beginning of both association and dissociation phases. These spikes can be removed for esthetic purposes or can be left in the sensorgrams, since these regions of the response curves are usually left out during the quantitative analysis. Sometimes for technical reasons spikes (air bubble or dust particle in sample or buffer) can appear during the experiment (Sect. 3.8). These sensorgram distortions can be removed or the data can be made invalid for analysis.

In the BIAevaluation software, all the described sensorgram operations can be carried out in a graphical way without number manipulations in the data spreadsheets.

4.1
1: 1 Interactions Analyzed by Quantitative Analysis

4.1.1
Introduction

The SPR technology was developed for extracting kinetic parameters from the interaction between macromolecules. In this part we will recapitulate the analysis theory for the simplest event involving binding of a single protein molecule to a single binding site on DNA. A more extensive description of the binding theory can be found in [11–16].

As described in Sect. 2.2, the response curve can be divided into association (including equilibrium) and dissociation phases (Fig. 3). The kinetic parameters can be extracted from both phases.

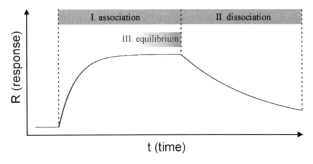

Fig. 3 The three parts of a response curve used in data analysis

4.1.2
Association

During the injection phase of an SPR experiment, termed association, the kinetic processes of $DNA \cdot PROTEIN$ complex formation and its decay can be represented by the scheme:

$$DNA + PROTEIN \underset{k_{off}}{\overset{k_{on}}{\rightleftharpoons}} DNA \cdot PROTEIN \ .$$

The rate of the $DNA \cdot PROTEIN$ complex formation is expressed as:

$$\frac{d(DNA \cdot PROTEIN)}{dt} = k_{on}[DNA][PROTEIN] - k_{off}[DNA \cdot PROTEIN] \ .$$

where [DNA] and [PROTEIN] are the concentrations of free or unbound DNA and protein, respectively. Assuming that DNA is immobilized at the chip surface (Sect. 3.2) whereas protein is injected, this equation can be written using SPR terminology as:

$$\frac{d(R)}{dt} = k_{on}(R_{max} - R) \cdot C - k_{off} \cdot R \ . \tag{1}$$

where R is the response, R_{max} is the response at DNA saturation by protein, and C is the concentration of injected protein (the amount of protein bound to DNA is negligible).

Taking the natural logarithm of Eq. 1

$$\ln \frac{d(R)}{dt} = \ln(k_{on} \cdot R_{max} \cdot C) - (k_{on} \cdot C + k_{off}) \cdot t \ ,$$

and substituting

$$k_{obs} = k_{on} \cdot C + k_{off} \ , \tag{2}$$

results in the equation:

$$\ln \frac{d(R)}{dt} = -k_{obs} \cdot t + \ln(k_{on} \cdot R_{max} \cdot C) \ . \tag{3}$$

Since for a given protein concentration C, k_{obs} and $\ln(k_{on} \cdot R_{max} \cdot C)$ are constant, Eq. 2 represents a linear function (Fig. 4a). Nonlinearity of the $\ln \frac{d(R)}{dt} = f(t)$ plot suggests that the interaction cannot be analyzed using a 1 : 1 model or that other processes, such as mass transport (Sect. 3.3), interfere with $DNA \cdot PROTEIN$ complex formation. Besides being an important diagnostic plot, the $\ln \frac{d(R)}{dt} = f(t)$ function can also be used for evaluation of the $DNA \cdot PROTEIN$ complex formation rate constant k_{on}. The plot k_{obs} vs C is linear (Eq. 2) and the slope of this function represents the k_{on} rate constant (Fig. 4b). The intercept of $k_{obs} = k_{on} \cdot C + k_{off}$ corresponds to the k_{off} rate constant. However, this method for evaluation of k_{off} is not reliable since a small change in k_{on} results in a large alteration of k_{off}.

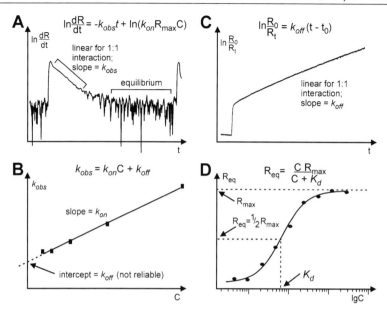

Fig. 4 Plots used during quantitative data analysis. See text for details

4.1.3
Dissociation

During the dissociation phase of an experiment, the protein solution is replaced by buffer and the response change represents solely the decay of the *DNA · PROTEIN* complex.

$$DNA \cdot PROTEIN \xrightarrow{k_{off}} DNA + PROTEIN \, .$$

The rate of this process is expressed as:

$$\frac{d(R)}{dt} = -k_{off} \cdot R \, . \tag{4}$$

The solution of this differential equation is:

$$R(t) = R_0 \cdot e^{-k_{off} \cdot t} \, .$$

This can be used for evaluation of the k_{off} rate constant by fitting to a single exponential or it can be converted to equation:

$$\ln \frac{R_0}{R_t} = k_{off}(t - t_0) \, , \tag{5}$$

where R_0 is the response at the analysis start point t_0 (which is not necessary at the beginning of the dissociation phase).

The slope of this function represents the rate constant k_{off} (Fig. 4c). A nonlinearity of the plot in Eq. 5 indicates deviation from the 1:1 model. Equation 5 does not include a protein concentration factor; a protein concentration dependent change in the slope of this plot suggests rebinding of protein to the DNA.

4.1.4
Equilibrium

This phase of the experiment can be analyzed as a special case of the association phase when the rate of $DNA \cdot PROTEIN$ complex formation equals the rate of its decay. By definition, at this point no change of response is observed:

$$\frac{d(R)}{dt} = 0.$$

At equilibrium Eq. 1 can be written as:

$$0 = k_{on} \cdot (R_{max} - R_{eq}) \cdot C - k_{off} \cdot R_{eq}$$

Rearranging for R_{eq} and dividing by k_{on} results in:

$$R_{eq} = \frac{C \cdot R_{max}}{C + \frac{k_{off}}{k_{on}}}.$$

Finally, substituting dissociation constant $K_d = \frac{k_{off}}{k_{on}}$ produces the equation:

$$R_{eq} = \frac{C \cdot R_{max}}{C + K_d}. \tag{6}$$

Equation 6 can be used for evaluation of the dissociation constant K_d, since K_d is the protein concentration at which $R_{eq} = 1/2 \cdot R_{max}$ (Fig. 4d). Equation 6 can be rearranged, especially in the case where saturation is not reached, using the Scatchard plot:

$$\frac{R_{eq}}{C} = -\frac{1}{K_d} R_{eq} + \frac{1}{K_d} R_{max}, \tag{7}$$

where the slope of the linear function $\frac{R_{eq}}{C} = f(R_{eq})$ represents $-\frac{1}{K_d}$.

4.1.5
Practical Guideline on Evaluation of the Rate Constants k_{on} and k_{off} and the Dissociation Constant K_d

The most popular SPR instrument, BIAcore, provides software which globally (simultaneously for the association and dissociation phases and for several protein concentrations) fits the response curves to Eqs. 1 and 4 using a nonlinear least-squares algorithm. Besides being displayed, the evaluated kinetic parameters are used for simulation. The simulated response curves are automatically overlaid with experimental data, and residuals plots (a plot representing the difference between measured curves and simulated curves) are generated. Often data analysis is recognized as credible when the value of χ^2 is lower than 2. This criterion is only rarely met when working with low amounts of immobilized DNA due to high noise, but the analysis can still be correct. Similarly, a bad study can fulfill very often the $\chi^2 < 2$ condition. For this reason careful examination of the residuals plots is strongly recommended. The experimental points should be randomly distributed around the fit, without any systematic deviations. Additionally to the residuals plots, researchers must check if the obtained kinetic parameters simply make sense, e.g., in respect of preliminary data obtained by gel-shift or footprint experiments (Sect. 3.1).

Additionally to the global analysis, the BIAevalution program also allows estimation of kinetic parameters separately from the dissociation and association phases (Eqs. 5, 2, and 3, respectively, Fig. 4a–c).

For most SPR applications the dissociation constant K_d is calculated from the $K_d = k_{off}/k_{on}$ proportion. It can also be evaluated from the steady-state phase (equilibrium). This method, however, is applicable only for kinetically fast interactions. The time required for reaching the equilibrium at a protein concentration equal to K_d can be expressed as $1/k_{off}$. For an interaction with $k_{off} = 1 \times 10^{-3}$ s^{-1}, it will take approximately 1 h to reach equilibrium. The steady state is reached faster with increased protein concentrations. Using a high protein concentration results, however, in the collection of only a narrow range of R_{eq} points, close to saturation (R_{max}). In this situation the K_d is

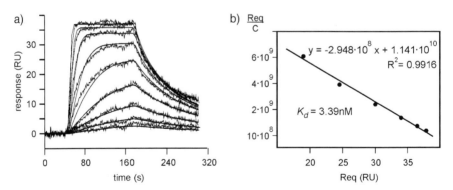

Fig. 5 SPR analysis of *E. coli* DnaA protein interaction with DNA containing one DnaA box. 12 RU of 21mer DNA was immobilized on an SA chip. The protein was injected at concentrations of 0.39, 0.78, 1.56, 3.13, 6.25, 12.5, 25, 50, and 100 nM at a flow rate of 100 µl min–1. The association phase (130 s) was followed by 120 s dissociation phase. **a** The experimental data (*rough lines*) and response curves simulation based on evaluated rate constants (*smooth lines*). **b** Scatchard equilibrium analysis for the six highest concentrations of DnaA from **a**

calculated as $1/2\, R_{max} = K_d$ (Fig. 4d) or from the slope of the Scatchard plot (Eq. 7 and Fig. 5b).

Figure 5 shows an example of data analysis using the 1 : 1 binding model, applied for binding of *Escherichia coli* DnaA protein to DNA containing a single binding site (the DnaA box from the promoter region of *dnaA* gene) (Speck and Messer, unpublished data). The kinetic rate constants $k_{on} = 8.09 \times 10^6$ ($M^{-1}s^{-1}$) and $k_{off} = 0.0262$ (s^{-1}) were evaluated by global analysis of binding curves for nine concentrations (0.39–100 nM) of DnaA (Fig. 5a). Values for the equilibrium response of the six highest protein concentrations were also used for calculation of the dissociation constant $K_d = 3.39$ nM using a Scatchard plot (Fig. 5b). Values of dissociation constants calculated from rate constants $K_d = 3.24$ nM and from steady-state $K_d = 3.39$ nM are almost identical. This criterion, as well as the simulation of the response based on the evaluated rate constant (smooth lines in Fig. 5a), validates the analysis.

4.2
Kinetic Analysis of Interactions Between Protein and a DNA Containing Multiple Protein Binding Sites

4.2.1
Introduction

Contrary to gel retardation or analytical ultracentrifugation techniques, where every *DNA · PROTEIN* species can be separated and quantitatively analyzed, the response curve in SPR experiments represents the sum of all binding events between protein and DNA containing multiple binding sites. For this reason quantitative analysis of protein binding to more then one binding site on DNA is relatively difficult, and successful only when binding events differ significantly in kinetic constants.

4.2.2
Curve Fitting with the BIAevaluation Software

The BIAevaluation software contains a preprogrammed model for studying interactions between an analyte and a heterogeneous ligand containing two binding sites. In this model the protein binds either site independently (non-cooperative, parallel reactions) as represented by the scheme:

$$DNA_1 + PROTEIN \underset{k_{off1}}{\overset{k_{on1}}{\rightleftarrows}} DNA_1 \cdot PROTEIN \quad \text{and}$$

$$DNA_2 + PROTEIN \underset{k_{off2}}{\overset{k_{on2}}{\rightleftarrows}} DNA_2 \cdot PROTEIN \,.$$

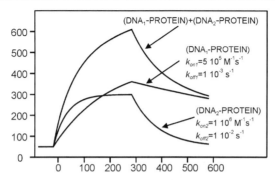

Fig. 6 Response curves simulation for interaction between protein and DNA containing two different and independent binding sites (parallel reactions)

To evaluate kinetic constants the data are fitted to a set of four differential equations:

$$\frac{d[DNA_1]}{dt} = -(k_{on1}[DNA_1][PROTEIN] - k_{off1}[DNA_1 \cdot PROTEIN])$$

$$\frac{d[DNA_1 \cdot PROTEIN]}{dt} = k_{on1}[DNA_1][PROTEIN] - k_{off1}[DNA_1 \cdot PROTEIN]$$

$$\frac{d[DNA_2]}{dt} = -(k_{on2}[DNA_2][PROTEIN] - k_{off2}[DNA_2 \cdot PROTEIN])$$

$$\frac{d[DNA_2 \cdot PROTEIN]}{dt} = k_{on2}[DNA_2][PROTEIN] - k_{off2}[DNA_2 \cdot PROTEIN].$$

Figure 6 shows simulated binding curves for protein binding to DNA containing two binding sites. Protein binds to both sites independently with an affinity of 2 and 10 nM for DNA binding site No. 1 and 2, respectively, but with different kinetics.

4.2.3
Hill's Cooperativity Factor

One area where using SPR to study protein binding to DNA with multiple binding sites can be very useful is in estimation of Hill's cooperativity factor. According to Hill's theory [17, 18], the simultaneous binding to DNA with n identical binding sites can be represented as:

$$h \cdot PROTEIN + DNA \overset{K_d^h}{\leftrightarrow} (PROTEIN_h \cdot DNA),$$

where K_d is the intrinsic dissociation constant for a single DNA binding site. The Hill coefficient h describes cooperativity as follows: $0 < h < 1$ indicates negative cooperativity, whereas $1 < h < n$ indicates positive cooperativity. n is the number of protein binding sites on the DNA. At equilibrium, the fraction of saturation Y (the ratio of occupied sites versus total sites) can be expressed as the function of free or unbound protein and K_d:

$$Y = \frac{[PROTEIN_{free}]^h}{(K_d + [PROTEIN_{free}]^h)}. \tag{8}$$

After rearrangement and assuming that the concentration of free protein equals the concentration of injected protein C, Eq. 8 can written as:

$$\log \frac{Y}{(1-Y)} = h \cdot \log C. \tag{9}$$

After expressing the saturation fraction in the SPR terminology $Y = R_{eq}/R_{max}$, the final form of Eq. 9 is:

$$\log \frac{R_{eq}/R_{max}}{(1 - R_{eq}/R_{max})} = h \cdot \log C, \tag{10}$$

where h is the slope of the function

$$\log \frac{R_{eq}/R_{max}}{(1 - R_{eq}/R_{max})} = f(\log C).$$

For a successful analysis the response value at the equilibrium R_{eq} for each protein concentration must be known. This requires long protein–DNA contact times, which can be achieved by injecting protein solutions at a slow flow rate (mass transfer does not influence the R_{eq} value). The second important parameter for analysis is the response value at protein saturation R_{max}. It can be difficult to reach, especially for negatively cooperative interaction. In this case the theoretical maximum binding value can be used:

$$R_{max_{cal}} = n \frac{R_{DNA} MW_{Protein}}{MW_{DNA}}, \tag{11}$$

where

R_{DNA} = amount of immobilized DNA (RU)
MW_{DNA} = molecular weight of DNA
$MW_{Protein}$ = molecular weight of protein
n = number of protein binding sites on DNA.

Even a 20% error of R_{max} estimation does not significantly influence the value of

$$\log \frac{R_{eq}/R_{max}}{(1 - R_{eq}/R_{max})},$$

especially for a large R_{eq}–R_{max} difference.

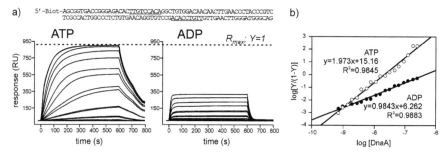

Fig. 7 Hill coefficient evaluation for an interaction between *S. coelicolor* DnaA and DNA containing two binding sites (TTGTCCACA) in a head-to-head orientation. The protein complexes with ATP (*panel A left*) or ADP (*panel A right*) were injected at a flow rate of 10 µl min^{-1} for 10 min followed by 3 min dissociation. DnaA was used at concentrations of 0.78, 0.89, 1.41, 1.8, 2.8, 3.6, 5.6, 7.2, 11.3, 14.4, 22.5, 28.8, 45.0, 57.5, 90.0, 115, 180, and 230 nM. **a** Experimental curves and DNA used. **b** Hill's analysis for equilibrium response values in **a**. The R_{max} for DnaA-ATP was used for both DnaA-ATP and DnaA-ADP

Figure 7 shows the ATP-induced cooperative binding of DnaA protein from *Streptomyces coelicolor* to DNA containing two DnaA binding sites (Majka and Messer, unpublished data). The ADP form of the protein interacts with DNA in a noncooperative manner, $h = 0.98$, whereas its ATP form displays an almost maximal value of positive cooperativity, $h = 1.97$. Since for ADP-DnaA saturation was not reached, the R_{max} value for ATP-DnaA was used for analysis of both DnaA forms.

A similar analysis was carried out to determine the Hill coefficient for the interaction between ethR repressor and *ethA* operator [19] (Sect. 6).

4.2.4
Cooperativity Fold α Factor

A different approach was used to decipher cooperativity during interaction of EST-1 transcription factor with a palindromic head-to-head ETS binding site in the stromelysin-1 promoter [20]. First the K_d of ETS binding to both palindrome half-sites was determined using a 1 : 1 model. Next the interaction between ETS and the complete palindrome was analyzed by using four different models: (1) two independent and nonequivalent binding sites on DNA, (2) two independent binding sites on DNA and conformational change, (3) sequential binding, and finally (4) sequential binding followed by conformational change.

Data were fitted to sets of differential equations describing each of the models and rate constants were used to calculate dissociation constants (up to three). For each model these constants were then used to calculate the apparent binding affinity of a single palindrome half-site of a full palindrome as the juxtaposition of two independent and equivalent binding sites:

$$K_{d,app}^{WT} = \sqrt[n]{\prod_{i=1}^{n} K_{di}}$$

The apparent affinity was then used to calculate the cooperativity fold α, expressed as:

$$\alpha = \frac{K_d^{M1} \cdot K_d^{M2}}{(K_{d,app}^{WT})^2},$$

where K_d^{M1} and K_d^{M2} are dissociation constants for the half-sites of the palindrome.

The cooperativity fold α (physically understood as by how many times the affinity of the cooperative interaction is larger than a hypothetical, noncooperative binding event) measured by this method was calculated as 8.8 and 19.7 for sequential binding and sequential binding followed by a conformational change model, respectively. Authors of the study found that these two models give the best description of the interaction.

5
Qualitative Analysis

The major application of SPR methods in studying DNA–protein binding is extracting the kinetic parameters of an interaction. However, qualitative analysis of the sensorgram can also provide useful information.

5.1
Stoichiometry Determination

The response measured in SPR experiments is not only proportional to the amount of protein–DNA bound to a chip surface (a prerequisite of kinetic analysis), but also the change in the bulk refractive index per unit change in protein concentration (specific refractive index increment) is closely similar for a wide range of proteins and nucleic acids. In other words, an equal mass of two different proteins (or protein and DNA) bound to the chip surface will give the same response value. The SPR technology can therefore be used for estimation of the DNA–protein complex stoichiometry according to the equation:

$$n = R_{max} \frac{MW_{DNA}}{R_{DNA} MW_{Protein}},$$

where

n = the number of protein molecules bound to DNA
= the number of protein binding sites on DNA
(assuming that a single protein molecule binds
a single binding site)
R_{max} = response for saturating concentration of protein
R_{DNA} = amount of immobilized DNA (RU) and
$MW_{DNA}, MW_{Protein}$ = molecular weight of DNA and protein, respectively.

5.2
Qualitative Analysis of Protein–DNA Complex Formation

Qualitative BIAcore analysis is very easy, fast, and requires only small amounts of protein. It is an ideal analytical tool to answer the question: does protein X bind to DNA y or DNA z or both, not more not less? Therefore, BIAcore analysis can be used for fast screening of protein mutants [21]. A variation of this theme is studying the influence of other factors on DNA–protein interaction. For example, BIAcore analysis revealed that two protein complexes, checkpoint clamp and its loader, interact with DNA only when mixed together and that ATP/Mg^{2+} is required for this reaction. Neither checkpoint clamp nor its loader alone can bind DNA, regardless of the presence of ATP/Mg^{2+} [22].

Cooperative interactions are frequently very complicated, therefore investigators often prefer qualitative over quantitative analysis. For example, plasminogen activator inhibitor-1 promoter (PAI-1) contains binding sites for two transcription regulators: Smads and the bHLH class of transcription factors TFE-3 and E47 [23]. The authors of the study investigated interaction between PAI-1 promoter and mixtures of Smad4/TFE-3 or Smad4/E47 and compared the response curves to those obtained for Smad4, TFE-3, and E47 run separately. They observed that the response curve for Smad4/E47 was identical to the sum of the curves obtained for Smad4 and E47, whereas the response curve for the Smad4/TFE-3 showed stronger interaction than the sum of curves Smad4 and TFE-3, thus indicating cooperative binding.

6
Surface Plasmon Resonance to Study Protein–DNA Interactions 1993–2006

The use of SPR to study protein–DNA interactions started in 1993 when Bondeson et al. reported the lactose repressor–operator interaction with the first commercially available BIAcore/SPR system [24]. Since then researchers have studied many different aspects of protein–DNA interactions using SPR; in the following we will review some studies of exceptionally high quality or creativity.

A typical example of a 1 : 1 protein–DNA interaction is the binding of the bacterial DNA-replication initiator DnaA to its 9-bp DNA recognition sequence, the DnaA box (Speck and Messer, unpublished data). This is illustrated in Fig. 5. Data were analyzed using the 1 : 1 binding model. This model, which contains only a minimal number of fitting parameters, describes the sensorgrams for the whole range of DnaA concentrations during both kinetic and equilibrium parts of the experiment very well. In addition, a binding stoichiometry close to 1 : 1 was determined indicating that the correct binding model was used and that the obtained kinetic parameters were reliable. Interestingly, similar data have been obtained with crude bacterial extracts containing overexpressed DnaA protein [21]. The authors used extract with overexpressed DnaA to characterize the DNA binding potential of a large number of point mutants which all resided in the DNA binding domain of DnaA. Interestingly, the mutants displayed changes in association and dissociation rates and in addition some had altered sequence specificity.

Biology frequently utilizes protein–protein interactions to regulate processes on DNA, e.g., transcription and DNA replication. Due to the ability of SPR to measure kinetics as well as the stoichiometry of complexes, a number of high quality studies have dissected the network of protein–protein and protein–DNA interactions and defined cooperativity in complex formation. Cooperativity can originate from protein–protein interactions, which can help a single protein to multimerize on DNA. This is the case for ethR, a bacterial repressor of transcription belonging to the TetR/CamR family of repressors [19]. The authors showed that ethR binds to the ethA promoter by gel shifts and DNase I footprinting. From these experiments they deduced the relative affinity of ethR for DNA and identified a specific binding region within the ethA promoter. Interestingly, in the center of the binding region they found two inverted repeats which themselves contain two imperfect inverted repeats. Next they analyzed the protein–protein interactions of ethR and found by gel filtration that ethR forms a dimer in solution. With this information they were able to design meaningful SPR experiments using protein concentrations deduced from the gel-shift experiments. The authors used a 55-bp DNA fragment of the ethA promoter spanning the region protected in the DNase I footprint + 3 bp on either end. Then equilibrium binding experiments were performed using long injection times to reach maximum binding levels. The data were drawn in a Hill plot and a Hill coefficient of 3.45 ($n = 4$) indicated very strong positive cooperativity in complex formation (Sect. 4.2.3). Stoichiometry analysis revealed an 8 : 1 ethR/DNA stoichiometry, fitting with two perfect and two imperfect inverted repeats as binding sites for four ethR dimers. A subset of binding sites spanning one perfect repeat and two imperfect repeats was bound very inefficiently; however, two perfect repeats and one imperfect repeat in combination were bound very well leading to a 4.5 : 1 EthR/DNA complex. These data indicate that binding to two perfect repeats enables cooperative binding of the perfect and imperfect repeat, likely through protein–protein interactions between the ethR dimers. In respect of transcriptional regulation the cooperativity allows a rapid regulatory change of ethA expression with a minimal variation in ethR protein concentration.

In another typical example of cooperativity, two proteins with low sequence specificity interact to form a unit with increased sequence specificity and higher affinity. This scenario of cooperativity is frequently found in eukaryotic transcriptional regulation. For example, Grinberg and Kerppola reported that TFE3 and Smad4, two proteins with low sequence specificity, cooperate through protein–protein interactions and for that reason bind very efficiently to the plasminogen activator inhibitor-1 (PAI-1) promotor, which results in activation of PAI-1 transcription [23]. Interestingly, binding between a close homologue of TFE3, E47, and Smad4 was not cooperative, probably due to the lack of a protein–protein interaction. This indicates that a transcription factor with low affinity or specificity for DNA needs an interaction partner to bind efficiently to DNA and to result in a change of transcriptional activity.

A number of groups have used SPR to study the formation of large protein–DNA complexes, containing several different proteins bound to DNA. The focus of this research is not on the determination of binding constants but on qualitative terms (i.e., does formation of the complex depend on subunit x,y,z or does ATP binding and hydrolysis influence complex formation). Pacek et al. showed recruitment of the bacterial DNA helicase DnaB by DnaA and DnaC [25]. They immobilized a 64-bp DNA fragment originating from the plasmid RK2 on a sensor chip. Then they injected DnaA onto this surface and analyzed binding of DnaC, DnaB, and DnaB–DnaC to the DnaA–DNA complex. The DnaB–DnaC complex bound very efficiently to the DnaA–DNA complex; however, neither DnaB nor DnaC individually bound the DnaA–DNA complex. In addition, they showed that the plasmid initiator protein TrfA can interact with a DNA–DnaA–DnaB–DnaC complex in a DnaB–DnaC dependent fashion. In another study, also analyzing complex assembly in qualitative terms, Majka and Burgers reported on RFC-Rad24, a five-subunit complex, and the DNA damage checkpoint sliding clamp Rad17/Mec3/Ddc1, which form cooperatively a complex on DNA in an ATP dependent fashion (see Sect. 5.2 for details) [22]. These two studies underscore that quantitative analysis of huge DNA–protein complexes can yield very informative data, which are difficult to obtain with other techniques.

Finally, we want to report briefly on some interesting assays or screens which are based on protein–DNA interactions. Hao et al. adopted the BIAcore for rapid determination of sequence specificity of DNA binding proteins (BIAcore-based SELEX assay) [26]. They immobilized the histidine-tagged DNA binding domain of a transcription factor and injected a pool of randomized oligomers. DNA that was bound efficiently was amplified by PCR and injected again. After several cycles they sequenced 33 clones and obtained a consensus sequence for the transcription factor.

Maesawa et al. developed an assay to measure telomerase rate using the BIAcore to detect elongation [27]. They analyzed crude samples from normal and cancerous tissue and found increased telomerase activity in several cancer tissues. They suggest that this method has advantages in clinical research over conventional techniques.

Lastly, the BIAcore has also been used to study proteins that bind nonspecifically to DNA [28], peptide–DNA interactions [29], and the generation of artificial DNA binding proteins [30–32].

7
Technological Developments

BIAcore introduced the first commercial product to study biomolecular interactions with surface plasmon resonance in 1990. Since then the technology has developed rapidly, resulting in the huge variety of instruments available today (Sect. 2.2). Sensitivity has increased dramatically, resulting in a product specifically suited to studying small-molecule interactions (BIAcore S51). However, demand for higher sensitivity will continue in order to detect weak small-molecule interactions, low levels of biomolecules present in crude samples, and to further reduce the problem of mass transport in biomolecular interaction. Toward this aim Hu et al. reported recently the development of an Au nanocluster embedded in a dielectric film to achieve a tenfold improvement in sensitivity [33].

Presently SPR is used to study individual protein–DNA interactions with respect to kinetics, thermodynamics [5], or stoichiometry. One important development in the future will be to detect simultaneously a large number of interactions in parallel. BIAcore recognized this demand and recently introduced the Flexchip, which uses a sensor chip that allows simultaneous analysis of 400 interactions on a single biosensor, however with a slightly reduced sensitivity. Two studies, using alternative SPR detection systems (SPR imaging), have recently shown that the analysis of protein–DNA interactions on a large scale is feasible [34, 35]. Systems biology will profit from the multiplicity of this technology to generate interaction maps, which are based on kinetic data rather than qualitative data.

Acknowledgements We thank C. Weigel and T. Mazurek for critical reading of the manuscript.

References

1. Watson JD, Crick FH (1953) Nature 171:737
2. Galas DJ, Schmitz A (1978) Nucleic Acids Res 5:3157
3. Carey J (1991) Methods Enzymol 208:103
4. Rich RL, Myszka DG (2005) J Mol Recognit 18:431
5. Oda M, Nakamura H (2000) Genes Cells 5:319
6. Myszka DG, He X, Dembo M, Morton TA, Goldstein B (1998) Biophys J 75:583
7. Hart DJ, Speight RE, Cooper MA, Sutherland JD, Blackburn JM (1999) Nucleic Acids Res 27:1063
8. Cui X, Yang F, Li A, Yang X (2005) Anal Biochem 342:173
9. Myszka DG (2000) Methods Enzymol 323:325
10. Myszka DG (1999) J Mol Recognit 12:279
11. BIAcore AB (1997) BIAevaluation version 3.0 software handbook
12. BIAcore AB (1998) BIAtechnology handbook
13. O'Shannessy DJ, Brigham-Burke M, Soneson KK, Hensley P, Brooks I (1993) Anal Biochem 212:457
14. O'Shannessy DJ, Winzor DJ (1996) Anal Biochem 236:275
15. Bjorquist P, Bostrom S (1997) Thromb Res 85:225
16. Jonsson U, Fagerstam L, Ivarsson B, Johnsson B, Karlsson R, Lundh K, Lofas S, Persson B, Roos H, Ronnberg I et al (1991) Biotechniques 11:620
17. Cornish-Bowden A (1979) Fundamentals of enzyme kinetics. Butterworth, London & Boston
18. Fersht AR (1985) Cooperative ligand binding, allosteric interactions, and regulation. Freeman, New York
19. Engohang-Ndong J, Baillat D, Aumercier M, Bellefontaine F, Besra GS, Locht C, Baulard AR (2004) Mol Microbiol 51:175
20. Baillat D, Begue A, Stehelin D, Aumercier M (2002) J Biol Chem 277:29386
21. Blaesing F, Weigel C, Welzeck M, Messer W (2000) Mol Microbiol 36:557
22. Majka J, Burgers PM (2003) Proc Natl Acad Sci USA 100:2249
23. Grinberg AV, Kerppola T (2003) J Biol Chem 278:11227
24. Bondeson K, Frostell-Karlsson A, Fagerstam L, Magnusson G (1993) Anal Biochem 214:245
25. Pacek M, Konopa G, Konieczny I (2001) J Biol Chem 276:23639
26. Hao D, Ohme-Takagi M, Yamasaki K (2003) FEBS Lett 536:151
27. Maesawa C, Inaba T, Sato H, Iijima S, Ishida K, Terashima M, Sato R, Suzuki M, Yashima A, Ogasawara S, Oikawa H, Sato N, Saito K, Masuda T (2003) Nucleic Acids Res 31:E4
28. Yoshioka K, Saito K, Tanabe T, Yamamoto A, Ando Y, Nakamura Y, Shirakawa H, Yoshida M (1999) Biochemistry 38:589
29. Kubo T, Yokoyama K, Ueki R, Abe S, Goto K, Niidome T, Aoyagi H, Iwakuma K, Ando S, Ono S, Fujii M (2000) Nucleic Acids Symp Ser 49
30. Imanishi M, Sugiura Y (2002) Biochemistry 41:1328
31. Nomura W, Sugiura Y (2003) Biochemistry 42:14805
32. Oda M, Furukawa K, Sarai A, Nakamura H (1999) Biochem Biophys Res Commun 262:94
33. Hu WP, Chen SJ, Huang KT, Hsu JH, Chen WY, Chang GL, Lai KA (2004) Biosens Bioelectron 19:1465
34. Maillart E, Brengel-Pesce K, Capela D, Roget A, Livache T, Canva M, Levy Y, Soussi T (2004) Oncogene 23:5543
35. Wegner GJ, Lee HJ, Marriott G, Corn RM (2003) Anal Chem 75:4740

Adv Biochem Engin/Biotechnol (2007) 104: 37–64
DOI 10.1007/10_027
© Springer-Verlag Berlin Heidelberg 2006
Published online: 8 September 2006

Identification of Regulatory Elements by Gene Family Footprinting and In Vivo Analysis

David F. Fischer[1,2] · Claude Backendorf[1] (✉)

[1]Laboratory of Molecular Genetics, Leiden Institute of Chemistry, Leiden University,
P.O. Box 9502, 2300 RA Leiden, The Netherlands
backendo@chem.leidenuniv.nl

[2]*Present address:*
Galapagos Genomics, Archimedesweg 4, 2333CN Leiden, The Netherlands

1	Identification of Transcription Factor Binding Sites	38
2	Evolution Guides the Identification of Regulatory Elements	38
2.1	Phylogenetic Footprints	38
2.2	Differential Phylogenetic Footprinting	39
3	Gene Family Footprinting	41
3.1	Paralogous Genes Can Diverge in Coding Potential and Expression Pattern	41
3.2	Molecular Changes in Paralogous Gene Regulation	42
3.3	Recently Duplicated Genes	43
3.4	The Epidermal Differentiation Complex (EDC)	43
3.5	CE Precursor Genes and "Fused Genes"	44
4	Proof-of-Principle: Novel Elements in the *SPRR3* Promoter	50
4.1	Identification of *SPRR3*-Specific Regulatory Elements in the Proximal Promoter	51
4.2	Binding Specificity of the Two *SPRR3*-Specific Complexes	51
4.3	*SPRR3*-Specific Complex C (CDTF-1) Requires Calcium for DNA Binding	55
5	Conclusions	57
References		59

Abstract Gene families of recently duplicated but subsequently diverged genes provide an unique opportunity for comparative analysis of regulatory elements. We have studied the human *SPRR* gene family of small proline rich proteins involved in barrier function of stratified squamous epithelia. These genes are all expressed in normal human keratinocytes, but respond differently to environmental insults. Comparisons of the functional promoter regions allows the rapid identification of both conserved and of novel regulatory elements that appeared after gene duplication. Competitive electrophoretic mobility shift assays can be used to confirm their presence.
  Here we show the power of gene family footprinting by the identification of two novel elements in the *SPRR3* promoter, not present in *SPRR1A* and *SPRR2A*. One of these elements binds a protein similar to GAAP-1, a pro-apoptotic activator of IRF-1 and p53. In vivo analysis shows that this element functions as an inhibitor of *SPRR3* transcription. The second novel element functions as an activator of promoter activity and is characterized by its A/T rich sequence. The latter interacting protein indeed binds through

contacts in the minor groove, and strikingly, depends on the presence of calcium for DNA interaction.

Keywords Gene family · Transcription factor · Keratinocyte · Transfection · Terminal differentiation · Promoter

1
Identification of Transcription Factor Binding Sites

DNA-protein interactions provide the framework for the regulation of gene expression. The identification of transcription factor binding sites is commonly done by (1) the use of bioinformatics to predict potential sites in the genome [1, 2], (2) in vitro experiments to confirm binding of these factors to DNA [3], and (3) in vivo validation of the role of the transcription factor in the regulation of the target gene. Here we will review the use of evolutionary information for the rapid identification of conserved and novel regulatory elements, and show in vivo validation of this approach from the *SPRR* gene family as a proof-of-principle.

2
Evolution Guides the Identification of Regulatory Elements

Generally, the conservation of the sequence of a given gene after divergence of two or more species reveals nucleotides under selection to preserve function. The comparison of such orthologous genes can either reveal that function and expression have remained identical during evolution, or that a change has occurred along one or more lineages. In the former case, comparison of the promoter region of the genes will yield a phylogenetic footprint [4, 5]. Only those sequences on which negative selection is active, i.e., the *cis*-elements, will be conserved over longer periods of time. If, on the other hand, a change in expression pattern has occurred during evolution, differential phylogenetic footprinting can be employed to reveal the molecular factors responsible for the difference [6]. In this case, conserved sequences are most likely not involved in the difference in gene expression between the species, whereas base changes might be indicative for novel regulatory elements.

2.1
Phylogenetic Footprints

In the first report on phylogenetic footprinting [4], ε-globin and γ-globin sequences from 10 species were compared, which yielded several regions of high sequence conservation over 70 Myr. A number of these regions has subsequently been shown to contain *cis*-elements involved in the expression of

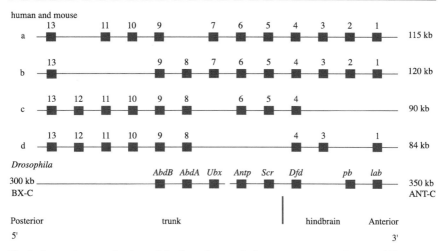

Fig. 1 Genomic organization of the *hox* clusters in human, mouse and *Drosophila*. Genes are indicated by *black boxes*, the size in kilobases of the clusters is indicated. Note that the *Drosophila* complexes BX-C and ANT-C are not physically linked. The expression pattern is indicated underneath. Data is from [12, 50, 177, 178]

these genes [7]. Significantly one conserved region (− 114 in the γ-globin promoter) was not only identified by phylogenetic footprinting, but also by a disease-causing mutation [8]. Conserved regions identified by phylogenetic footprinting have been analyzed for functional conservation. Deletion of the regions in the mouse *hoxb4* enhancer which are conserved in the pufferfish *Fugu rubripes* (speciation occurred approximately 430 Myr ago) showed that these are indeed important *cis*-elements, moreover *Fugu* sequences could drive *hoxb4*-specific expression in transgenic mice [9]. An enhancer shared by *hoxb3* and *hoxb4* contains two *cis*-elements conserved in human, mouse, chicken and pufferfish [10]. This bidirectional enhancer has not been identified in *Drosophila*, which has been explained by the larger distances between genes in the two complexes (BX-C and ANT-C [10–12](Fig. 1)). Phylogenetic footprinting thus has its limitations: comparison of sequences from closely related organisms will not display enough mutations outside of *cis*-elements, whereas sequences from organisms separated too far in evolution will not reveal any residual homology.

2.2
Differential Phylogenetic Footprinting

During evolution, orthologous genes can acquire a different expression pattern, for instance in the β-globin locus, human γ-globin is expressed in the fetus, whereas in mouse, rabbit and prosimian primates, γ-globin is expressed in the embryo [4, 13] (Fig. 2). Analysis of the 5′-flanking regions of

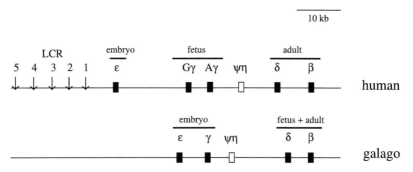

Fig. 2 Genomic organization of the β-globin locus from human [179] and the prosimian *Galago crassicaudatus* [4, 181]. Genes are indicated by *black boxes*, pseudogenes by *open boxes*; the locus control region (LCR, [181]) is indicated. The expression pattern is indicated above each gene

the γ-globin gene from human and from *Galago crassicaudatus* (the speciation of simian primates and strepsirhine primates occurred approximately 55 Myr ago) revealed several nucleotide changes [4, 6, 14, 15]. Using several pairs of oligonucleotides from the human and *Galago* promoters, harboring these polymorphisms, different protein binding patterns were indeed identified [6, 16]. However only one of these nucleotide changes was subsequently found to affect promoter strength [6], which could indicate that the other binding sites are redundant (e.g., [17]). Transgenic mouse studies have nevertheless shown that changes in *cis*-elements are the cause for the evolution of fetal expression: a 4 kilobase fragment of the human γ-globin was highly expressed in mouse fetal liver, whereas the comparable fragment from the *Galago* γ-globin gene was expressed selectively in embryonic life [18]. Differential methylation has been implicated in the regulation of expression in the human β-globin cluster: expressed genes are hypomethylated [19], furthermore, forced demethylation can reactivate γ-globin expression in vivo [20]. A hypomethylated human-specific element from the γ-globin promoter was bound by the stage selector protein, implicated in fetal expression [16]. Methylation of this site induced competition by Sp1, which could be responsible for the down-regulation of this gene in adult life [21]. The stage selector protein does not bind to the *Galago* γ-globin promoter [16], suggesting that the acquisition of this protein to globin gene regulation is specific for simians.

Similar experiments have been performed with the mouse and chicken *hoxc8* genes (speciation occurred approximately 220 Myr ago). Mice have 7 cervical and 13 thoracic vertebrae, whereas chicken have 14 cervical and 7 thoracic vertebrae [22]. The expression of *hoxc8* in somites and the neural tube has shifted similarly along the anterior-posterior axis [23]. A 399 bp fragment from the mouse *hoxc8* early enhancer has been shown to direct tissue-specific expression in transgenic mice [24, 25]. Swapping of a 151 bp fragment with the chicken *hoxc8* early enhancer shifted the expression pattern posteriorly, simulating a chicken relational pattern [26]. As the 151 bp fragment is still considerably conserved between chicken and mouse (80%), identification of diverged regulatory elements has been partially performed [27].

Expression levels and androgen inducibility of the *D7Rp2e* gene are highly variable among species of the genus *Mus* [28]. Differential phylogenetic footprinting of seven mouse species revealed several mutations in transcription factor binding sites, resulting in a difference in DNA binding affinity of identical factors present in diverged mouse species [29]. Swapping experiments have indeed shown that these binding sites are the determinant for the different expression levels of *D7Rp2e* in *Mus domesticus* and *Mus pahari* [30], two species which have diverged approximately 8–10 Myr ago. In conclusion, a change in gene expression can be obtained by several means: a binding site can become non-functional by mutation, it can acquire a different function by the same mechanism, or can arise de novo. Even more subtle changes (relative affinity) can result in a different expression pattern of orthologous genes in two species.

3
Gene Family Footprinting

3.1
Paralogous Genes Can Diverge in Coding Potential and Expression Pattern

Gene duplication generally results in the creation of a redundant copy of the locus. A redundant copy is not under the negative selection for detrimental mutations, and can thus evolve at a high rate to a new gene and can spread more easily in the population [31, 32]. Genome sequencing reveals that local clusters of duplicated genes are highly abundant, e.g., the *Caenorhabditis elegans* genome harbors 402 of these gene families with up to 20 genes in each cluster [33]. In the human genome, over 1500 conserved gene families have been identified [34]. Members of a multigene family that have the same function are believed to undergo homogenization by concerted evolution [35, 36]. A good example of concerted evolution is the pair of γ-globin genes in the β-globin locus: allelic genes differ only slightly, whereas non-allelic genes differ considerably, indicating that intergenic gene conversion homogenizes these genes [37]. Alternatively to concerted evolution, the two genes can diverge in structure and expression subsequent to the duplication event. Both processes occur only slowly, which results in multigene families whose individual members show considerable homology and similar expression patterns [31]. Gene replacement experiments can elucidate whether differences in expression pattern or differences in protein structure are important for the function of each member of a multigene family.

Keratins are members of the superfamily of intermediate filament proteins, important for cellular structure (reviewed in [38]). The different keratin genes have similar, but distinct expression patterns [39]. For example mitotically active cells of many stratified epithelia express keratins K5 and K14, whereas terminally differentiating keratinocytes in these epithelia express keratins K1 and K10 [40]. Simple epithelia express among others keratins K8 and K18 [39]. The question whether keratins K14 and K18 are equivalent in function has been resolved by transgenic mice in which the K14 coding sequence was replaced by the K18 coding sequence [41]. These mice rescue some of the phenotypes associated with a K14 knockout [42], but they are deficient in withstanding mechanical stress [41]. Thus, specific protein structures are important in these members of the keratin gene family. The keratin K6 genes on the other hand are expressed by multiple genes in human [43] and mouse [44]. The paralogous K6 proteins are more similar (95% and 97%) than the orthologous proteins (maximally 82%) [44]. Comparison of the nucleotide sequences however showed that the proximal promoter sequences of the orthologous human K6α and mouse K6α are highly similar, whereas the K6β sequences and the paralogous sequences are substantially less conserved. These findings suggest that the transcriptional regulation of the K6α genes is important; indeed the two mouse K6 genes are differentially expressed, whereas all amino acid differences are expected to be conservative [44]. The function of the different K6 genes is thus probably most easily addressed by mutating the promoter regions of the paralogues.

The mouse *En-1* and *En-2* proteins (55% amino acid identity) are related to the *Drosophila* segmentation gene *engrailed*, and exhibit diverged expression patterns [45]. The phenotypes of mice nullizygous for either gene are different: for example $En\text{-}1^{-/-}$ is lethal, whereas $En\text{-}2^{-/-}$ mice are viable; the $En\text{-}1^{-/-}$ phenotype can however be completely rescued by replacing *En-1* coding sequences with *En-2* coding sequences [46]. This indicates that the functional difference between *En-1* and *En-2* is caused by their divergent expression patterns. Similarly the three paired-box and homeobox proteins *paired* (*prd*), *gooseberry* (*gsb*) and *gooseberry neuro* (*gsbn*) have diverged considerably [47], and also the expression patterns for these genes are distinct [48]. The finding that *prd* coding sequences can replace *gsb* coding sequences to rescue a *gsb*⁻ phenotype indicates that the expression patterns of these genes are important for their function rather than their coding sequences [49].

In conclusion, paralogous genes can acquire different functions by changing their expression pattern or transcriptional responsiveness, whereas the coding sequences can remain functionally identical.

3.2
Molecular Changes in Paralogous Gene Regulation

As discussed above, evolutionary changes can only be readily observed in a limited time-window: separated genes (either by speciation or duplication) should have accumulated enough mutations to result in differential expression or to permit phylogenetic footprinting. Longer times will however result in a severe change in regulation, elucidation of which is beyond the simple analysis of sequences. This situation has for instance occurred in the *hox* gene cluster: the duplication events that have generated the paralogous genes in the different clusters (Fig. 1) are likely to predate the speciation of pseudocoelomates (nematodes) and coelomates (insects and chordates) [11, 50]. The first duplication event (cognate groups 1 to 7 versus groups 9 to 13) has been postulated to have occurred 1000 Myr ago [50], which infers that all subsequent duplications have occurred between 1000 and 550 Myr [51]. In order to perform comparative analysis of paralogous genes, younger gene families are thus preferred.

3.3
Recently Duplicated Genes

Two examples from *Drosophila* species demonstrate how gross rearrangements can result in novel genes with novel expression patterns. The *jingwei* gene from *D. teissieri* and *D. yakuba* was created in a common ancestor (approximately 17 to 20 Myr ago) by retrotransposition of the alcohol dehydrogenase gene into the third intron of the *yande* gene [52]. The *jingwei* gene has thus obtained an expression pattern different from the parental *Adh* gene. The function of *jingwei* gene has also changed, and *JGW* encodes an alcohol dehydrogenase with altered substrate specificity [53].

In *D. melanogaster*, a newly (approximately 3 Myr ago) evolved dynein gene is proposed to originate from the fusion of a duplicated cell-adhesion protein annexin X (AnnX) gene and a duplicated gene encoding a cytoplasmic dynein intermediate chain (*Cdic*) [54]. The resulting gene (*Sdic*) has not only obtained novel protein coding sequences, but also a novel, sperm-specific, expression pattern. The novel *Sdic* promoter appears to have been created largely through serendipity by juxtaposition of sequences that strongly resemble testis-specific promoter elements, even though they originated from non-regulatory sequences [54]. The sperm-specific expression pattern has been postulated to have assisted in positive selection of mutations in this gene, and indeed this protein may function as an axonemal dynein intermediate chain in sperm [55].

3.4
The Epidermal Differentiation Complex (EDC)

Human chromosome 1 reveals in region 1q21 a most remarkable density of genes termed epidermal differentiation complex (EDC) [56], that fulfill important functions in terminal differentiation of the human epidermis [57–59]. These genes encode (1) the cornified cell envelope (CE) precursors, including loricrin [60, 61], involucrin [62], the 11 small proline rich repeat

(*SPRR*) proteins [59, 63, 64] and the 21 members of the LCE family of late cornified envelope proteins [65], (2) at least 16 S100 calcium-binding proteins (*S100A1-S100A16*) [58, 66] and (3) the "fused genes"/intermediate filament associated proteins (IFAP) profilaggrin [67, 68], trichohyalin [69, 70] repetin [71], hornerin [72] and cornulin [73].

Sequence analysis revealed that these genes are likely to descent from either an ancestral CE precursor gene or *S100* gene [56, 74], the fused genes combine domains of both CE precursor genes and *S100* genes [71, 75-78]. The human epidermal differentiation complex has been mapped to a 2.3 Mbase region [79, 80]; for most of these genes, mouse orthologues, which are also linked, have been identified on chromosome 3 [66, 71, 77, 78, 81-87]. This indicates that although the different duplication events giving rise to the paralogues of the CE precursor family, S100A family and fused gene family have occurred at least 80 Myr ago, these genes have remained physically linked. Because of their physical linkage, concerted evolution could occur between different paralogues [35, 36], maintaining protein structure and/or expression [59].

Besides their similar genomic organization, these genes are all expressed in squamous epithelia (Table 1). The S100 genes are also differentially expressed in non-epithelial tissues [88], whereas the CE precursors and IFAPs are more-or-less restricted to squamous epithelia, although in these tissues, their expression patterns can differ [64, 71, 89-92]. Strikingly, expression of *SPRRs* has recently been reported in numerous non-squamous and even non-epithelial tissues [93-97], mainly upon tissue injury, inferring that *SPRRs* might have additional protective functions (e.g., protection against oxidative stress [97]). These findings suggest that the paralogues should display at least some conserved regulatory elements involved in epithelial (epidermal) expression, but should also reveal specific elements for each gene-family or gene.

3.5
CE Precursor Genes and "Fused Genes"

The small proline rich proteins constitute a specific sub-class of cornified cell envelope (CE) precursors [77], encoded by a multigene family clustered within the epidermal differentiation complex (EDC) at human chromosome 1q21 [56, 59]. The locus contains two *SPRR1* genes, seven *SPRR2* genes, a single *SPRR3* gene and a single *SPRR4* gene [64, 114]. Analysis of the regions required for expression of these CE precursor genes and the three best-studied fused genes has revealed that the most important regulatory elements are confined to the promoter regions (Table 1). The introns of involucrin and *SPRR2A* have been reported to enhance transcription of their promoters, but the introns do not confer keratinocyte-specific or differentiation-specific signals [110, 112, 116]. Sequence comparison of the CE precursor and fused gene promoter regions shows that several paralogues display significant similarity: homology exists between the *SPRR1* and *SPRR3* genes, between the *SPRR2A* and *SPRR2C* genes and between the profilaggrin (*FLG*), trichohyalin (*THH*) and repetin (*RPTN*) genes (data not shown). No conservation can be observed between other paralogues, although loricrin and involucrin mouse orthologues do show a high degree of conservation with their human counterparts. Significantly, the minimal promoter region required for expression of the profilaggrin and the trichohyalin gene has been mapped to position –116 and position –135, respectively [77, 103]. No data is available yet on the requirements for the repetin gene [71, 148], although Klf4 has been shown to regulate mouse repetin expression [149]. The homology between the three best-studied fused gene promoters is confined to this proximal promoter region (Fig. 3). Besides a conserved TATA box, also the important AP-1 site identified in the profilaggrin gene is conserved between the three promoters (note that the core AP-1 recognition site is a palindrome [150], thus the A to T transversion in *RPTN* is conservative). The Ets binding site of *FLG* [106] is conserved in the *THH*, but not in the *RPTN* gene. Consistent with the observed conservation of regulatory elements in the *FLG* and *THH* genes, co-expression of these proteins has been observed [151]. Subtle differences in gene expression do however exist: for instance normal human keratinocytes induced to terminally differentiate by a shift in calcium concentration induce both filaggrin and trichohyalin, but expression of the former occurs at an earlier time point [89]. A remarkable difference between the two promoters, a 17 bp deletion in the *FLG* promoter between the AP-1 site and the TATA box (Fig. 3), could be related to this phenomenon, as such deletions have been shown to affect the responsiveness of a given gene (see above, *ADH1*, and below, *SPRR2A*).

The *SPRR1* and *SPRR3* genes show a significant homology in their promoter sequences (Fig. 4). The human *SPRR1A* and *SPRR3* promoters have

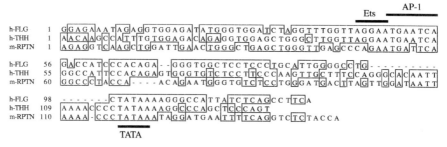

Fig. 3 Comparison of the proximal promoter sequences of the fused genes. The proximal promoter sequences of human profilaggrin (*FLG*), human trichohyalin (*THH*) and mouse repetin (*RPTN*) were aligned with the SeqVu program (The Garvan Institute for Medical Research, version 1.1). The TATA box, and Ets and AP-1 sites in the *FLG* gene [105] are indicated

been studied extensively [92, 118] and the important regulatory elements are indicated in Fig. 4. Conserved in each of the four genes are the TATA box and zinc finger binding site, which are also found in the *SPRR2* and involucrin genes [64], indicating that these elements predate mouse-human speciation (approximately 80 Myr ago) and even predate the duplication events giving rise to the different CE precursors. The proximal Ets binding site conserved in the human *SPRR1* and *SPRR2* genes [92, 122, 152], is not present in the human *SPRR3* and mouse *SPRR1B* gene, which can be attributed to an independent single base mutation in either gene (Fig. 4) [118]. In the involucrin and loricrin genes, no Ets binding site has been identified [108, 123], indicating that this element was not present in the ancestral CE precursor gene, but was present in the ancestral *SPRR* gene [59]. The identification of the Ets sites at a different position (and in the opposite orientation) in the *FLG* and *THH* genes, suggests that these have independently evolved. This assumption is confirmed by the analysis of the human *SPRR3* gene, which has a distal Ets site instead of a proximal one [118]. Figure 4 shows that this distal Ets site is also present in the human *SPRR1B* gene, which could thus have two functional Ets sites. Although we have shown that a distal Ets site is not func-

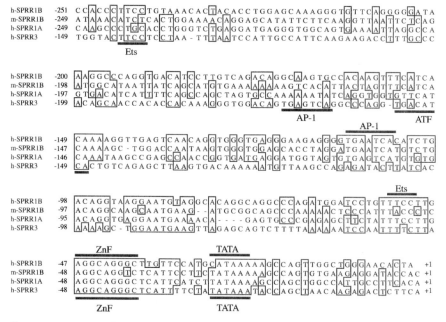

Fig. 4 Comparison of the proximal promoter sequences of the *SPRR1* and *SPRR3* genes. Human *SPRR1A*, *SPRR1B* and *SPRR3*, and mouse *SPRR1B* were aligned with the SeqVu program (The Garvan Institute for Medical Research, version 1.1). The TATA box, and transcription factor binding sites in the *SPRR1A* (*lines above the sequence*) and *SPRR3* (*lines underneath the sequence*) genes are indicated

Table 1 Genes in the epidermal differentiation complex (EDC)

Gene	Exons (coding)	Squamous epithelial expression (a)	Non-epithelial expression (b)	Minimal promoter (c)	Cis-elements (d)
Centromere					
S100A10	3 (2&3)	[98]	Brain [99]		
S100A11	3 (2&3)	[98, 100]	Muscle [101]		
THH (trichohyalin)	3 (2&3)	[76, 89, 102]		−135 to ? [103]	Ets, AP-1 [76]
RPTN (repetin)	3 (2&3)	[71]			
HRNR (hornerin)	3 (2&3)	[72]			
FLG (profilaggrin)	3 (2&3)	[68, 104]		−116 to +9 [105]	Ets [106], **AP-1** [105]
CRNN (cornulin)	3 (2&3)	[73]			AP-1 [73]
LCE genes (late cornified envelope precursors)	2 (2)	[65]			
IVL (involucrin)	2 (2)	[91, 107]		−2216 to −7 [108]; −2473 to +4400 [109] −2473 to +40 [110] −2473 to +1240 [111] −2473 to +156 [112] −2473 to +4000 [113]	**AP-1** [108]

Table 1 continued

Gene	Exons (coding)	Squamous epithelial expression (a)	Non-epithelial expression (b)	Minimal promoter (c)	Cis-elements (d)
SPRR4	2 (2)	[64, 114]	Heart (ESTs)		
SPRR1A	2 (2)	[64, 92]	Injured nerves [96, 115]	−125 to +42 [116]	**AP-1, Ets, zinc finger** [92]
SPRR3	2 (2)	[64, 117]		−1415 to +100	**Ets, AP-1, ATF, zinc finger** [118]
SPRR1B	2 (2)	[64, 119, 120]			**AP-1**, Ets, zinc finger
SPRR2D	2 (2)	[64, 119]			AP-1, octamer, Ets, zinc finger [64]
SPRR2A	2 (2)	[64, 121, 122]		−134 to +14	AP-1, **octamer, ISRE, Ets, zinc finger** [122]
SPRR2B	2 (2)	[64]			Octamer, Ets, zinc finger [64]
SPRR2E	2 (2)	[64, 119]			AP-1, octamer, Ets, zinc finger [64]
SPRR2F	2 (2)	[64]			AP-1, octamer, zinc finger [64]
SPRR2C	2 (2)	[64, 119]			AP-1, octamer, Ets, zinc finger [64]
SPRR2G	2 (2)	[64]			AP-1, Ets, zinc finger [64]
LOR (loricrin)	2 (2)	[60, 91]		−60 to +1160 [123]	**AP-1** [123]
S100A9	3 (2&3)	[124, 125]	Myeloid [84, 126]	−1000 to +420 [126]	

Table 1 continued

Gene	Exons (coding)	Squamous epithelial expression (a)	Non-epithelial expression (b)	Minimal promoter (c)	Cis-elements (d)
S100A12	3 (2&3)	[125]	Myeloid [127]		
S100A8	3 (2&3)	[124, 125]	Myeloid [84, 126]	−1500 to +536 [126, 128]	
S100A7	3 (2&3)	[100, 125, 129, 130]		−167 to +134 [135],	
S100A6	3 (2&3)	[131, 132]	Muscle [133]	−361 to +134 [135]	
S100A5	4 (2-4)	[136]	Brain, muscle [136]		
S100A4	3 (2&3)	[131, 137]	Lymphoid, mesenchyme [138, 139]	−41 to +1254 [140–142]	**NF-κB** [141]
S100A3	3 (2&3)	[89, 131]	muscle [136]		
S100A2	3 (2&3)	[131, 143, 144]	Muscle [145]	−1336 to +17 [146]	AP-1 [146]
S100A13	3 (2&3)	[66]	Muscle, lymphoid, brain [66]		
S100A1 telomere	3 (2&3)	[143]	Brain, muscle [147]	−249 to +10 147	

The order of the genes in the epidermal differentiation complex is as given in the table from centromere to telomere (*top to bottom*) (see http://www.ensembl.org) [56, 79, 80]. **a** Evidence for expression in squamous epithelial tissues or cultured epidermal keratinocytes. **b** Evidence for expression in non-epithelial tissues. **c** The minimal regulatory region required for proper expression of the gene in transgenic mice or transfected tissue culture. **d** Regulatory elements identified by homology or DNA-binding activity (*roman letterface*) or identified as essential (≥ 50% reduction of activity) for promoter activity (*bold face*)

tionally identical to a proximal Ets site [118], a general enhancement of gene expression by the distal Ets site can be expected in the case of the *SPRR1B* gene, which is expressed in human keratinocytes at a significantly higher level than the *SPRR1A* gene [92]. In line with the model of evolutionary "tinkering" [153], the *SPRR1/3* ancestral gene is likely to have had two Ets sequences (as in *SPRR1B*), subsequently, *SPRR3* has lost the proximal Ets site, *SPRR1A*, has lost the distal Ets site, whereas mouse *SPRR1B* (which is poorly expressed as compared to its *SPRR1A* paralogue [86, 154]) has apparently lost both sites.

Conserved in the *SPRR*/involucrin/loricrin class of CE precursor genes and fused genes is an AP-1 binding site [73, 92, 105, 108, 118, 120, 121, 123, 148, 155] (Table 1). In most cases where the functionality of this sequence in keratinocyte gene expression has been tested (*SPRR1A, SPRR1B, SPRR3, IVL, LOR, FLG*), mutation resulted in a severe reduction of promoter activity (Table 1). In the *SPRR2A* gene however, this sequence has little importance during early stages of keratinocyte terminal differentiation [122].

A loss-of-function of a given *cis*-element has been identified for the *SPRR3* Ets site: although the same Ets transcription factor (ESX/ESE-1) [156, 157] transactivates via both the proximal *SPRR1A/SPRR2A* Ets site and via the distal *SPRR3* Ets site, this transcription factor is involved in the TPA-induction of the *SPRR1A* gene, but apparently does not confer TPA-responsiveness to the *SPRR3* gene [92, 118] (unpublished data). The absence of a highly synergistic transcription complex on the *SPRR3* promoter [118], in contrast to the *SPRR1A* and *SPRR2A* promoters [92, 122], is likely to have allowed the addition of several novel regulatory elements to the *SPRR3* gene [118]. Which of these changes is responsible for the unique expression pattern of *SPRR3* [117, 158, 159], remains to be elucidated (see also below). The acquisition of different expression patterns of the CE precursor genes is likely to have allowed the modulation of the biomechanical properties of epithelia in different tissues [78, 160, 161].

4
Proof-of-Principle: Novel Elements in the *SPRR3* Promoter

The *SPRR3* gene has diverged from the *SPRR1A* and *SPRR2A* genes in both structure [59, 118] and expression [117]. These genes are expressed in squamous epithelia [63, 162], and in in vitro cultured normal human keratinocytes, where terminal differentiation and *SPRR* expression can be modulated by the extracellular calcium concentration [92, 118, 121, 122]. Analysis of the promoter regions of these three genes has revealed several regulatory elements required for expression during keratinocyte terminal differentiation. One element common to all genes is an Ets binding site [92, 118, 122], recognized by the epithelium-specific transcription factor ESX/ESE-1 [156, 157]. Another element conserved in these three genes is a zinc finger binding site to which members of the Klf/Sp1 family bind [122]. In the *SPRR1A* and *SPRR2A* genes, these two binding sites are juxtaposed in the proximal promoter region, whereas in the *SPRR3* gene a single base mutation has disturbed the proximal Ets binding site [118]. Because the minimal promoter region required for expression of *SPRR1A* and *SPRR2A* does not extend beyond −125 and −134, respectively [92, 116, 122], clustering of regulatory elements close to the TATA box seems to be an important feature of these genes. Statistical analysis has revealed that several regulatory elements occupy only specific position in unrelated promoter sequences [163, 164]. Previous analysis of the *SPRR3* promoter has pointed to crucial elements at greater distance, but has not thoroughly examined the possibility that novel regulatory elements have evolved in the proximal *SPRR3* promoter after the evolutionary loss of the Ets binding site [118].

4.1
Identification of *SPRR3*-Specific Regulatory Elements in the Proximal Promoter

Comparison with *SPRR1A* and *SPRR2A*, mutational analysis and transient transfections in primary human keratinocytes indicated that additional regulatory elements are present in the sequences flanking the *SPRR3* zinc finger binding site (Fig. 5).

Competitive electrophoretic mobility shift assays [165] were performed to identify the proteins binding to the putative regulatory elements in the *SPRR1A*, *SPRR2A* and *SPRR3* proximal promoter regions. The *SPRR3* probe did not bind Ets proteins, whereas it did bind zinc finger proteins (Fig. 6). On the 40-mer probe one *SPRR3*-specific complex (P) could be detected, a slightly longer probe (47-mer, Fig. 7) revealed another *SPRR3*-specific complex (C). Factor C binds specifically to the upstream half (28-mer) of the 47-mer, whereas factor P binds to the downstream half (Fig. 7).

4.2
Binding Specificity of the Two *SPRR3*-Specific Complexes

As shown in Fig. 7, complex C required the presence of divalent cations (lane 2), whereas complex P did not; formation of complex C was strongest in the presence of calcium (see below). Neither of these two complexes resembled NF-Y (Fig. 7). Inspection of the sequence required for complex P formation revealed a CATTT core sequence (mutation in pSG433). The same motif was recently identified in both the *IRF-1* and *p53* genes, where it was shown to bind a 68 kDa pro-apoptotic nuclear factor, GAAP-1 [166, 167]. Competition experiments showed that *SPRR3*-specific factor P, but not factor C, binds to both the *p53* and *IRF-1* IPCS (Fig. 8). These data, and the apparent molecular weight of complex P as observed in an electrophoretic mobility shift assay

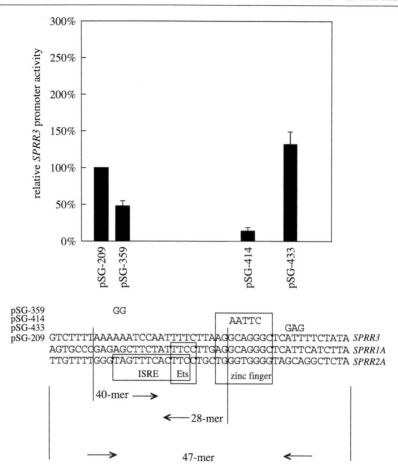

Fig. 5 Mutational analysis of the *SPRR3* proximal promoter region: normal human foreskin keratinocytes were cultured and transfected as described previously [122]. The wild type *SPRR3* promoter fused to the CAT reporter gene (pSG-209) was compared to mutant promoters (pSG-359, pSG-414 and pSG-433), which were made by mutagenesis [182] of the wild type *SPRR3* promoter [118]. CAT activity was measured 40 hr after induction of stratification according to [183] in four independent experiments and is expressed as the activity relative to wild type pSG-209. Mutations are indicated underneath the graph; the wild type *SPRR3* proximal promoter sequence (GenBank Acc. No. AF077374) is compared with the sequences from *SPRR1A* (GenBank Acc. No. L05187) and *SPRR2A* (GenBank Acc. No. X53064). The zinc finger binding site in the three genes is indicated by a *box*; the Ets binding sites in *SPRR1A* and *SPRR2A* are indicated, as is the ISRE (interferon stimulated responsive element) in *SPRR2A*. The 5′end of the 40-mer and the 3′end of the 28-mer are indicated in the figure; the 47-mer contained the complete sequence shown. As expected, the *SPRR3* Klf/Sp1 zinc finger binding site is a crucial regulatory element (mutated in pSG-414) [165]. Mutation of an (A)-rich sequence also resulted in a reduction of promoter activity in normal human keratinocytes (pSG-359), whereas mutation in a sequence downstream from the zinc finger binding site (pSG-433) resulted in an upregulation of promoter activity

Identification of Regulatory Elements by Gene Family Footprinting

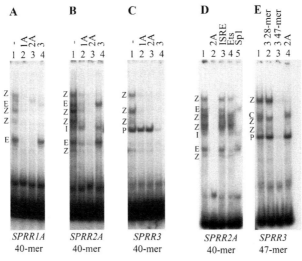

Fig. 6 Electrophoretic mobility shift assay of keratinocyte nuclear extracts with the proximal promoter regions of *SPRR1A* (A), *SPRR2A* (B and D) and *SPRR3* (C and E). Labeled probes are indicated underneath the figure and correspond to the sequences from the *SPRR1A*, *SPRR2A* and *SPRR3* promoters, represented in Fig. 5. The various competing oligonucleotides are indicated on top of the figure. The *SPRR1A* probe showed six complexes (panel A: lane 1), which were specific as they were competed for with excess unlabeled *SPRR1A* oligonucleotide (lane 2). Only one complex was not competed by the *SPRR2A* probe (lane 3), whereas two complexes were resistant to *SPRR3* competition (lane 4). Seven complexes were identified on the *SPRR2A* probe (panels A and D) and either four or five complexes on the *SPRR3* probes, depending on whether a 40-mer (panel C) or 47-mer probe (panel E) was used. The nature of the various complexes could be determined by competition with consensus oligonucleotides for the ISRE, Ets and zinc finger proteins. An example of such an analysis is shown in panel D: lane 1 no competitor DNA; lane 2: competition with *SPRR2A* 40-mer; lane 3: competition with a consensus ISRE; lane 4: competition with a consensus Ets binding site and lane 5: competition with a consensus Sp1 binding site. The identity of the various complexes is indicated at the left of each panel. Z: zinc finger protein (Sp1/Klf) binding site; E: Ets binding protein; I: ISRE binding protein. The two *SPRR3*-specific complexes are indicated with respectively C (CDTF-1) and P (IPCS). EMSAs of the *SPRR1A*, *SPRR2A* and *SPRR3* 40-mer proximal promoter oligonucleotides were essentially performed as previously described [122, 165]. An amount of 3 µg of keratinocyte nuclear extract was incubated for 5 min at RT in 20 µl of reaction buffer (10 mM HEPES-KOH pH 7.9, 50 mM NaCl, 10% glycerol, 0.5 mM EDTA, 0.5 mM ZnCl$_2$, 1 mM DTT, 50 µg/ml poly(dIdC)/(dIdC) (Pharmacia, Uppsala, Sweden), 250 µg/ml bovine serum albumin). The reaction buffer for the *SPRR3* 28-mer and *SPRR3* 47-mer oligonucleotides was identical to the previous buffer, except that ZnCl$_2$ was replaced by 1 mM CaCl$_2$. Subsequently, 20 fmol of [^{32}P]-labeled, double-stranded oligonucleotide) was added, and incubation was prolonged for 30 min at RT. Complexes were separated by electrophoresis at RT on a 4% polyacrylamide gel (60 : 1 molar ratio of acrylamide and bisacrylamide), containing 25 mM Tris-base, 190 mM glycine, 1 mM EDTA, 2.5% glycerol for 90 min at 10 V/cm. Competition experiments were performed by addition of an excess of oligonucleotide prior to addition of the labeled oligonucleotide. Oligonucleotide sequences are shown in Table 2

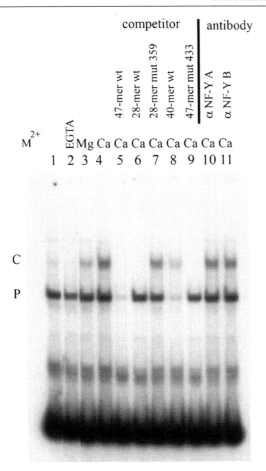

SPRR3 47-mer

Fig. 7 Analysis of factor C and P binding to the *SPRR3* promoter: keratinocyte nuclear extract was analyzed with the *SPRR3* 47-mer probe. In all lanes, a 250-fold excess of *SPRR2A* 40-mer was present to visualize only *SPRR3*-specific complexes. In lane 1, no divalent cations or chelators were added; in lane 2, 5 mM EGTA was added; in lane 3, 1 mM MgCl$_2$ was added; in lanes 4 to 9, 1 mM CaCl$_2$ was added. Lanes 5 to 9 contained a 250-fold excess of various *SPRR3* oligonucleotide competitor, indicated above the lanes. The wt 28-mer competed the C but not the P complex. A 28-mer containing the 359 mutation was no more able to compete the C complex. As the mutation in pSG-359 decreased *SPRR3* promoter activity by approximately 50%, it appears that complex C contains a transcriptional activator for *SPRR3*. Complex P was competed by the 47-mer (lane 4) and the 40-mer (lane 8). This competition was abolished by the mutation in pSG-433 (lane 9). Note that this mutation in pSG433 increased *SPRR3* promoter activity by approximately 1.5, indicating that complex P contains a weak transcriptional repressor. NF-Y has been shown to be the major CCAAT-binding activity in eukaryotic cells ([164], and references therein). Antibodies against two subunits of NF-Y however did not react with complexes C or P on the *SPRR3* probe (lanes 10 and 11). Oligonucleotide sequences are shown in Table 2

Identification of Regulatory Elements by Gene Family Footprinting

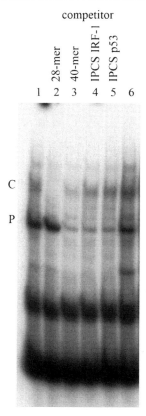

Fig. 8 Characterization of *SPRR3*-specific complex P. An electrophoretic mobility shift assay was performed under conditions identical to Fig. 7, lane 4. In lanes 1 and 6, no additional competitor oligonucleotide was present, in lanes 2 to 5 a 250-fold excess of the following competitors was added: lane 2 contained the *SPRR3* 28-mer, lane 3 the *SPRR3* 40-mer, lane 4 the IPCS/GAAP-1 binding site from *IRF-1* [166] and lane 5 the IPCS/GAAP-1 binding site from *p53*. The resulting EMSA indicated that both IPCS binding sites competed specifically the P but not the C complex. Oligonucleotide sequences are shown in Table 2

suggest that GAAP-1 and the *SPRR3*-specific factor P might be the same protein. Factor P appears to function as an inhibitor of *SPRR3* transcription (Fig. 5).

4.3
SPRR3-Specific Complex C (CDTF-1) Requires Calcium for DNA Binding

As shown in Fig. 7, *SPRR3*-specific complex C required divalent cations for DNA binding. Subsequent experiments showed that besides calcium, magnesium and manganese stimulated binding, but to a lesser extent, whereas zinc, copper, nickel and cobalt were detrimental to DNA binding activity (data not shown). The optimal Ca^{2+} concentration was determined to be 1 mM (data not shown). We therefore designated *SPRR3*-specific complex C: CDTF-1 (calcium dependent transcription factor). The CDTF-1 binding site was further studied in electrophoretic mobility shift assays. Mutational analysis identified the A/T rich core sequence TA$_6$T as the minimal binding site (Fig. 9).

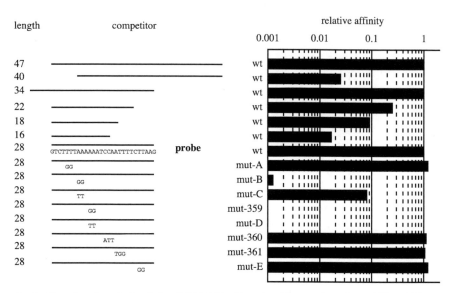

Fig. 9 Mutational analysis of the CDTF-1 binding site. Electrophoretic mobility shift assays were performed on the *SPRR3* 28-mer in the presence of 1 mM CaCl$_2$. For efficient DNA binding, a long oligonucleotide was required: deletion of 7 bases upstream of the (A/T) tract (40-mer) or 6 to 12 bases downstream of the (A/T) tract (22 to 16-mer) resulted in progressive drop in affinity. Addition of sequences (47-mer or 34-mer) to the 28-mer did however not result in an increase in activity, indicating that the 28-mer has the optimal length. Mutational analysis indicated that the (A) tract is crucial for DNA recognition. Whereas a mutant in the upstream thymine residues (mut-A) and mutants downstream of the (A) tract (mut-360, mut-361 and mut-E) did not affect binding activity, the mutants in the (A) tract (mut-B, mut-D and mut-359) severely reduced binding activity (by three orders of magnitude). Comparison of mut-359 and mut-D showed that the adenine residues cannot be replaced by thymine residues. Mut-C still retained some binding activity (one order of magnitude reduced), but this could be the result of a partial binding site inversion (CT$_4$A$_6$T to CT$_6$A$_4$T). The minimal binding site is thus TA$_6$T, although non-specific DNA contacts outside of this core sequence are required for optimal binding. Competition was performed with the oligonucleotides depicted in the figure at 5-fold, 25-fold, 125-fold, and 500-fold excess. Residual complex formation was quantified with a Betascope 603 blot analyzer (Betagen). The relative affinity is the ratio of wild type 28-mer competitor to mutant competitor required to achieve a 50% reduction in complex formation. Oligonucleotide sequences are shown in Table 2

Table 2 Oligonucleotides used in this study

Oligo	Gene	Sequence	Refs.
SPRR1A 40-mer	SPRR1A	gagagcttctatttccttgaggcagggctcattcatctta	[92]
SPRR2A 40-mer	SPRR2A	gggtagtttcacttcctgctgggtggggtagcaggctcta	[121]
SPRR3 40-mer	SPRR3	aaaaaatccaattttcttaaggcagggctcattttctata	[118]
Ets	polyomavirus	tcgagcaggaagttcgacgt	
Sp1	SV40 virus	aaatagtcccgcccctaactccgcccat	
47-mer	SPRR3	gtcttttaaaaaatccaattttcttaaggcagggctcattttctata	
47-mer mut 433	SPRR3	gtcttttaaaaaatccaattttcttaaggcagggctcgagttctata	
IPCS IRF-1	IRF-1	agcctgatttccccgaaatgacggcacgcagcc	[166]
IPCS p53	p53	aatgcaggattcctccaaaatgatttccac	[166]
wt 34-mer	SPRR3	agagcagtcttttaaaaaatccaattttcttaag	
28-mer	SPRR3	gtcttttaaaaaatccaattttcttaag	
wt 22-mer	SPRR3	gtcttttaaaaaatccaatttt	
wt 18-mer	SPRR3	gtcttttaaaaaatccaa	
wt 16-mer	SPRR3	gtcttttaaaaaatcc	
mut-A	SPRR3	gtctggtaaaaaatccaattttcttaag	
mut-B	SPRR3	gtcttttggaaaatccaattttcttaag	
mut-C	SPRR3	gtcttttttaaaatccaattttcttaag	
mut 359	SPRR3	gtcttttaaaggatccaattttcttaag	
mut-D	SPRR3	gtcttttaaattatccaattttcttaag	
mut-360	SPRR3	gtcttttaaaaaatattattttcttaag	
mut-361	SPRR3	gtcttttaaaaaatccatggttcttaag	
mut-E	SPRR3	gtcttttaaaaaatccaattttcggaag	

5
Conclusions

After divergence of the *SPRR3* gene from the *SPRR1* genes [59], a mutation in the proximal Ets binding site has disrupted DNA binding of ESE-1 [118], possibly allowing the creation of novel regulatory elements at this position in *SPRR3*. We have indeed found two *SPRR3*-specific regulatory elements in the proximal promoter region next to the conserved and functional zinc finger binding site. The element located between the zinc finger binding site and the TATA box has only weak, and negative transcriptional activity in normal human keratinocytes. It is conceivable that this element fulfills a more important role in the regulation of *SPRR3* in vivo, since *SPRR3* is expressed selectively in the esophagus, whereas *SPRR1* and *SPRR2* are expressed in the epidermis [117], yet in the in vitro culture system all three genes are expressed. This element was shown to bind a nuclear factor (P) similar to GAAP-1, which interacts with two recently regulatory elements in the *IRF-1* and *p53* genes [167]. In these genes, the regulatory element (IPCS) acted as an activator binding site involved in basal expression; it is not yet known whether this element also plays a role in terminal differentiation-linked expression, but GAAP-1 has been shown to possess pro-apoptotic activity [166, 167].

Upstream of the zinc finger binding site, an activator binding site was located in the *SPRR3* promoter (Fig. 5). The protein which bound specifically to this sequence, required divalent cations for DNA-binding. Our results suggest that the endogenous cation is calcium, since in vitro, optimal DNA-binding was observed with calcium; furthermore the calcium-selective chelator EGTA efficiently counteracted DNA-binding activity. The optimal calcium concentration for CDTF-1 we have observed is similar to the nuclear calcium concentration [Ca]n determined in various cell types (0.1 to 7 mM) [168–170], and furthermore the [Ca]n has been reported to be several fold higher than the [Mg]n [168, 171]. Calcium has been shown to act on transcription factors by binding with calmodulin or S100 proteins to basic helix-loop-helix proteins and thus inhibiting DNA-binding [172, 173]. Remarkably, the S100A1 to S100A16 proteins are also encoded by genes in the EDC at 1q21, where the *SPRR* genes are located [57, 58]. It is not unlikely that S100 proteins could also interact with other transcription factors, resulting in activation of DNA binding. The DNA sequence recognized by CDTF-1 is very (A/T) rich, which suggests that CDTF-1 interactions occur in the minor groove of the DNA. This was confirmed by the finding that Hoechst 33258, which is a minor groove-binding dye, reduced the DNA-binding activity of CDTF-1, whereas the same concentration of the intercalating drugs ethidium bromide and propidium iodide had no effect (data not shown). A similar behavior has been reported for the minor groove-binding protein TBP (TATA box binding protein) [174].

The isolation of a calcium dependent transcription factor for the *SPRR3* promoter is thought-provoking, since calcium ions plays an important role in the regulation of *SPRR* genes [118, 122] and keratinocyte terminal differentiation in vitro [175, 176]. Expression-analysis of GAAP-1 (complex-P) and molecular cloning of CDTF-1 should assist in the elucidation of the physiological roles these transcription factors might play during epidermal differentiation.

Comparisons of the functional *SPRR* promoter regions has allowed the rapid identification of both conserved and of novel regulatory elements that appeared after gene duplication. In vivo analysis has been used to indicate their functionality. Competitive electrophoretic mobility shift assays has confirmed their presence and aided in revealing their identity.

Acknowledgements We would like to thank Drs. R. Mantovani (Milan) and T. Taniguchi (Osaka) for the generous gift of antibodies and plasmids. This work was supported by the J.A. Cohen Institute, and by grants from NWO-SON and the EC.

References

1. Matys V, Kel-Margoulis OV, Fricke E, Liebich I, Land S, Barre-Dirrie A, Reuter I, Chekmenev D, Krull M, Hornischer K, Voss N, Stegmaier P, Lewicki-Potapov B, Saxel H, Kel AE, Wingender E (2006) Nucleic Acids Res 34:D108
2. Xie X, Lu J, Kulbokas EJ, Golub TR, Mootha V, Lindblad-Toh K, Lander ES, Kellis M (2005) Nature 434:338
3. Moss T (2001) DNA-protein interactions: principles and protocols. Humana Press, Totowa, NJ
4. Tagle DA, Koop BF, Goodman M, Slightom JL, Hess DL, Jones RT (1988) J Mol Biol 203:439
5. Gumucio DL, Shelton DA, Zhu W, Millinoff D, Gray T, Bock JH, Slightom JL, Goodman M (1996) Mol Phylogenet Evol 5:18
6. Gumucio DL, Shelton DA, Blanchard-McQuate K, Gray T, Tarle S, Heilstedt-Williamson H, Slightom JL, Collins F, Goodman M (1994) J Biol Chem 269:15371
7. Gumucio DL, Heilstedt-Williamson H, Gray TA, Tarle SA, Shelton DA, Tagle DA, Slightom JL, Goodman M, Collins FS (1992) Mol Cell Biol 12:4919
8. Fucharoen S, Shimizu K, Fukumaki Y (1990) Nucleic Acids Res 18:5245
9. Aparicio S, Morrison A, Gould A, Gilthorpe J, Chaudhuri C, Rigby P, Krumlauf R, Brenner S (1995) Proc Natl Acad Sci USA 92:1684
10. Gould A, Morrison A, Sproat G, White RA, Krumlauf R (1997) Genes Dev 11:900
11. Kappen C, Schughart K, Ruddle FH (1989) Proc Natl Acad Sci USA 86:5459
12. Ruddle FH, Bartels JL, Bentley KL, Kappen C, Murtha MT, Pendleton JW (1994) Annu Rev Genet 28:423
13. Goodman M, Koop BF, Czelusniak J, Weiss ML (1984) J Mol Biol 180:803
14. Hayasaka K, Fitch DH, Slightom JL, Goodman M (1992) J Mol Biol 224:875
15. Chiu CH, Schneider H, Slightom JL, Gumucio DL, Goodman M (1997) Gene 205:47
16. Jane SM, Ney PA, Vanin EF, Gumucio DL, Nienhuis AW (1992) EMBO J 11:2961
17. Gumucio DL, Shelton DA, Bailey WJ, Slightom JL, Goodman M (1993) Proc Natl Acad Sci USA 90:6018
18. TomHon C, Zhu W, Millinoff D, Hayasaka K, Slightom JL, Goodman M, Gumucio DL (1997) J Biol Chem 272:14062
19. Mavilio F, Giampaolo A, Carè A, Migliaccio G, Calandrini M, Russo G, Pagliardi GL, Mastroberardino G, Marinucci M, Peschle C (1983) Proc Natl Acad Sci USA 80:6907
20. Ley TJ, DeSimone J, Anagnou NP, Keller GH, Humphries RK, Turner PH, Young NS, Keller P, Nienhuis AW (1982) N Engl J Med 307:1469
21. Jane SM, Gumucio DL, Ney PA, Cunningham JM, Nienhuis AW (1993) Mol Cell Biol 13:3272
22. Carroll SB (1995) Nature 376:479
23. Burke AC, Nelson CE, Morgan BA, Tabin C (1995) Development 121:333
24. Shashikant CS, Bieberich CJ, Belting HG, Wang JC, Borbély MA, Ruddle FH (1995) Development 121:4339
25. Shashikant CS, Ruddle FH (1996) Proc Natl Acad Sci USA 93:12364
26. Belting HG, Shashikant CS, Ruddle FH (1998) Proc Natl Acad Sci USA 95:2355
27. Anand S, Wang WCH, Powell DR, Bolanowski SA, Zhang J, Ledje C, Pawashe AB, Amemiya CT, Shashikant CS (2003) Proc Natl Acad Sci USA 100:15666
28. Tseng-Crank J, Berger FG (1987) Genetics 116:593
29. Singh N, Barbour KW, Berger FG (1998) Mol Biol Evol 15:312
30. Singh N, Berger FG (1998) J Mol Evol 46:639

31. Ohno S (1970) Evolution by gene duplication. Springer, Berlin Heidelberg New York
32. Lynch M, Conery JS (2000) Science 290:1151
33. The *C. elegans* sequencing consortium (1998) Science 282:2012
34. Friedman R, Hughes AL (2003) Mol Biol Evol 20:154
35. Smith GP (1973) Cold Spring Harbor Symp Quant Biol 38:507
36. Dover GA (1986) Trends Genet 2:159
37. Slightom JL, Blechl AE, Smithies O (1980) Cell 21:627
38. Fuchs E (1995) Annu Rev Cell Dev Biol 11:123
39. Moll R, Franke WW, Sciller DL (1982) Cell 31:11
40. Fuchs E, Green H (1980) Cell 19:1033
41. Hutton E, Paladini RD, Yu QC, Yen M, Coulombe PA, Fuchs E (1998) J Cell Biol 143:487
42. Lloyd C, Yu QC, Cheng J, Turksen K, Degenstein L, Hutton E, Fuchs E (1995) J Cell Biol 129:1329
43. Takahashi K, Paladini RD, Coulombe PA (1995) J Biol Chem 270:18581
44. Takahashi K, Yan B, Yamanishi K, Imamura S, Coulombe PA (1998) Genomics 53:170
45. Davis CA, Joyner AL (1988) Genes Dev 2:1736
46. Hanks M, Wurst W, Anson-Cartwright L, Auerbach AB, Joyner AL (1995) Science 269:679
47. Bürglin TR (1994) In: Duboule D (ed) Guidebook to the homeobox genes. Oxford University Press, Oxford, p 73
48. Baumgartner S, Bopp D, Burri M, Noll M (1987) Genes Dev 1:1247
49. Li X, Noll M (1994) Nature 367:83
50. Zhang J, Nei M (1996) Genetics 142:295
51. Knoll AH (1992) Science 256:622
52. Long M, Langley CH (1993) Science 260:91
53. Zhang J, Dean AM, Brunet F, Long M (2004) Proc Natl Acad Sci USA 101:16246
54. Nurminsky DL, Nurminskaya MV, De Agular D, Hartl DL (1998) Nature 396:572
55. Ranz JM, Ponce AR, Hartl DL, Nurminsky D (2003) Genetica 118:233
56. Mischke D, Korge BP, Marenholz I, Volz A, Ziegler A (1996) J Invest Dermatol 106:989
57. Volz A, Korge BP, Compton JG, Ziegler A, Steinert PM, Mischke D (1993) Genomics 18:92
58. Schäfer BW, Wicki R, Engelkamp D, Mattei M-G, Heizmann CW (1995) Genomics 25:638
59. Gibbs S, Fijneman R, Wiegant J, Geurts van Kessel A, van de Putte P, Backendorf C (1993) Genomics 16:630
60. Mehrel T, Hohl D, Rothnagel JA, Longley M, Bundman D, Cheng C, Lichti U, Bisher ME, Steven AC, Steinert PM, Yuspa SH, Roop DR (1990) Cell 61:1103
61. Yoneda K, McBride OW, Korge BP, Kim IG, Steinert PM (1992) J Dermatol 19:761
62. Eckert RL, Green H (1986) Cell 46:583
63. Kartasova T, Muijen GNP, van Pelt-Heerschap H, van de Putte P (1988) Mol Cell Biol 8:2204
64. Cabral A, Voskamp P, Cleton-Jansen AM, South A, Nizetic D, Backendorf C (2001) J Biol Chem 276:19231
65. Jackson B, Tilli CL, Hardman M, Avilion A, Macleod M, Ashcroft G, Byrne C (2005) J Invest Dermatol 124:1062
66. Wicki R, Schäfer BW, Erne P, Heizmann CW (1996) Biochem Biophys Res Commun 227:594
67. Lonsdale-Eccles JD, Haugen JA, Dale BA (1980) J Biol Chem 255:2235

68. Presland RB, Haydock PV, Fleckman P, Nirunsuksiri W, Dale BA (1992) J Biol Chem 267:23772
69. O'Keefe EJ, Hamilton EH, Lee S-C, Steinert PM (1993) J Invest Dermatol 101:65S
70. Lee SC, Wang M, McBride OW, O'Keefe EJ, Kim IG, Steinert PM (1993) J Invest Dermatol 100:65
71. Krieg P, Schuppler M, Koesters R, Mincheva A, Lichter P, Marks F (1997) Genomics 43:339
72. Takaishi M, Makino T, Morohashi M, Huh NH (2005) J Biol Chem 280:4696
73. Contzler R, Favre B, Huber M, Hohl D (2005) J Invest Dermatol 124:990
74. Backendorf C, Hohl D (1992) Nat Genet 2:91
75. Markova NG, Marekov LN, Chipev CC, Gan S-Q, Idler WW, Steinert PM (1993) Mol Cell Biol 13:613
76. Lee S-C, Kim I-G, Marekov LN, O'Keefe EJ, Parry DAD, Steinert PM (1993) J Biol Chem 268:12164
77. Steinert PM, Marekov LN (1995) J Biol Chem 270:17702
78. Steinert PM, Kartasova T, Marekov LN (1998) J Biol Chem 273:11758
79. Marenholz I, Volz A, Ziegler A, Davies A, Ragoussis I, Korge BP, Mischke D (1996) Genomics 37:295
80. Lioumi M, Olavesen MG, Nizetic D, Ragoussis J (1998) Genomics 49:200
81. Rothnagel JA, Mehrel T, Idler WW, Roop DR, Steinert PM (1987) J Biol Chem 262:15643
82. Rothnagel JA, Longley MA, Bundman DS, Naylor SL, Lalley PA, Jenkins NA, Gilbert DJ, Copeland NG, Roop DR (1994) Genomics 23:450
83. Djian P, Phillips M, Easley K, Huang E, Simon M, Rice RH, Green H (1993) Mol Biol Evol 10:1136
84. Lagasse E, Weissman IL (1992) Blood 79:1907
85. Kizawa K, Tsuchimoto S, Hashimoto K, Uchiwa H (1998) J Invest Dermatol 111:879
86. Song H-J, Poy G, Darwiche N, Lichti U, Kuroki T, Steinert PM, Kartasova T (1999) Genomics 55:28
87. Ridinger K, Ilg EC, Niggli FK, Heizmann CW, Schäfer BW (1998) Biochim Biophys Acta 1448:254
88. Zimmer DB, Cornwall EH, Landar A, Song W (1995) Brain Res Bull 37:417
89. Ishida-Yamamoto A, Hashimoto Y, Manabe M, O'Guin WM, Dale BA, Iizuka H (1997) Br J Dermatol 137:9
90. Ishida-Yamamoto A, Kartasova T, Matsuo S, Kuroki T, Iizuka H (1997) J Invest Dermatol 108:12
91. Hohl D, Olano BR, de Viragh PA, Huber M, Detrisac CJ, Schnyder UW, Roop DR (1993) Differentiation 54:25
92. Sark MWJ, Fischer DF, de Meijer E, van de Putte P, Backendorf C (1998) J Biol Chem 273:24683
93. Nozaki I, Lunz JG 3rd, Specht S, Stolz DB, Taguchi K, Subbotin VM, Murase N, Demetris AJ (2005) Lab Invest 85:109
94. Morris JS, Stein T, Pringle MA, Davies CR, Weber-Hall S, Ferrier RK, Bell AK, Heath VJ, Gusterson BA (2006) J Cell Physiol 206:16
95. Carmichael ST, Archibeque I, Luke L, Nolan T, Momiy J, Li S (2005) Exp Neurol 193:291
96. Bonilla IE, Tanabe K, Strittmatter SM (2002) J Neurosci 22:1303
97. Pradervand S, Yasukawa H, Muller OG, Kjekshus H, Nakamura T, St Amand TR, Yajima T, Matsumura K, Duplain H, Iwatate M, Woodard S, Pedrazzini T, Ross J, Firsov D, Rossier BC, Hoshijima M, Chien KR (2004) Embo J 23:4517

98. Robinson NA, Lapic S, Welter JF, Eckert RL (1997) J Biol Chem 272:12035
99. De Leon M, Van Eldik LJ, Shooter EM (1991) J Neurosci Res 29:155
100. Moog-Lutz C, Bouillet P, Régnier CH, Tomasetto C, Mattei M-G, Chenard M-P, Anglard P, Rio M-C, Basset P (1995) Int J Cancer 63:297
101. Ohta H, Sasaki T, Naka M, Hiraoka O, Miyamoto C, Furuichi Y, Tanaka T (1991) FEBS Lett 295:93
102. Fietz MJ, Presland RB, Rogers GE (1990) J Cell Biol 110:427
103. Steinert P, Tarcsa E, Lee S-C, Jang S-I, Andreoli J, Markova N (1996) In: van Neste D, Randall VA (eds) Hair research for the next millenium. Elsevier Science, Amsterdam, p 169
104. Dale BA, Holbrook KA, Kimball JR, Hoff M, Sun T-T (1985) J Cell Biol 101:1257
105. Jang S-I, Steinert PM, Markova NG (1996) J Biol Chem 271:24105
106. Andreoli JM, Jang S-I, Chung E, Coticchia CM, Steinert PM, Markova NG (1997) Nucleic Acids Res 25:4287
107. Rice RH, Green H (1979) Cell 18:681
108. Welter JF, Crish JF, Agarwal C, Eckert RL (1995) J Biol Chem 270:12614
109. Crish JF, Howard JM, Zaim TM, Murthy S, Eckert RL (1993) Differentiation 53:191
110. Carroll JM, Taichman LB (1992) J Cell Sci 103:925
111. Carroll JM, Albers KM, Garlick JA, Harrington R, Taichman LB (1993) Proc Natl Acad Sci USA 90:10270
112. Ng DC, Su MJ, Kim R, Bikle DD (1996) Front Biosci 1:a16
113. Crish JF, Zaim TM, Eckert RL (1998) J Biol Chem 273:30460
114. Cabral A, Sayin A, de Winter S, Fischer DF, Pavel S, Backendorf C (2001) J Cell Sci 114:3837
115. Marklund N, Fulp CT, Shimizu S, Puri R, McMillan A, Strittmatter SM, McIntosh TK (2006) Exp Neurol 197:70
116. Fischer DF, van Drunen CM, Winkler GS, van de Putte P, Backendorf C (1998) Nucleic Acids Res 26:5288
117. Hohl D, de Viragh PA, Amiguet-Barras F, Gibbs S, Backendorf C, Huber M (1995) J Invest Dermatol 104:902
118. Fischer DF, Sark MWJ, Lehtola MM, Gibbs S, van de Putte P, Backendorf C (1999) Genomics 55:88
119. Kartasova T, van de Putte P (1988) Mol Cell Biol 8:2195
120. An G, Tesfaigzi J, Chuu Y-J, Wu R (1993) J Biol Chem 268:10977
121. Gibbs S, Lohman FP, Teubel W, Putte Pvd, Backendorf CMP (1990) Nucleic Acids Res 18:4401
122. Fischer DF, Gibbs S, van de Putte P, Backendorf C (1996) Mol Cell Biol 16:5365
123. DiSepio D, Jones A, Longley MA, Bundman D, Rothnagel JA, Roop DR (1995) J Biol Chem 270:10792
124. Saintigny G, Schmidt R, Shroot B, Juhlin L, Reichert U, Michels S (1992) J Invest Dermatol 99:639
125. Hitomi J, Kimura T, Kusumi E, Nakagawa S, Kuwabara S, Hatakeyama K, Yamaguchi K (1998) Arch Histol Cytol 61:163
126. Lagasse E, Clerc RG (1988) Mol Cell Biol 8:2402
127. Guignard F, Mauel J, Markert M (1995) Biochem J 309:395
128. Lagasse E, Weissman IL (1994) J Exp Med 179:1047
129. Hofmann HJ, Olsen E, Etzerodt M, Madsen P, Thøgersen HC, Kruse T, Celis JE (1994) J Invest Dermatol 103:370
130. Hardas BD, Zhao X, Zhang J, Longqing X, Stoll S, Elder JT (1996) J Invest Dermatol 106:753

131. Böni R, Burg G, Doguoglu A, Ilg EC, Schäfer BW, Müller B, Heizmann CW (1997) Br J Dermatol 137:39
132. Wood L, Carter D, Mills M, Hatzenbuhler N, Vogeli G (1991) J Invest Dermatol 96:383
133. Engelkamp D, Schäfer BW, Erne P, Heizmann CW (1992) Biochemistry 31:10258
134. Ferrari S, Calabretta B, deRiel JK, Battini R, Ghezzo F, Lauret E, Griffin C, Emanuel BS, Gurrieri F, Baserga R (1987) J Biol Chem 262:8325
135. van Groningen JJ, Weterman MA, Swart GW, Bloemers HP (1995) Biochem Biophys Res Commun 213:1122
136. Engelkamp D, Schäfer BW, Mattei MG, Erne P, Heizmann CW (1993) Proc Natl Acad Sci USA 90:6547
137. Shrestha P, Muramatsu Y, Kudeken W, Mori M, Takai Y, Ilg EC, Schäfer BW, Heizmann CW (1998) Virchows Arch 432:53
138. Ebralidze A, Tulchinsky E, Grigorian M, Afanasyeva A, Senin V, Revazova E, Lukanidin E (1989) Genes Dev 3:1086
139. Klingelhöfer J, Ambartsumian NS, Lukanidin EM (1997) Dev Dyn 210:87
140. Tulchinsky E, Kramerov D, Ford HL, Reshetnyak E, Lukanidin E, Zain S (1993) Oncogene 8:79
141. Tulchinsky E, Prokhortchouk E, Georgiev G, Lukanidin E (1997) J Biol Chem 272:4828
142. Chen D, Davies MP, Rudland PS, Barraclough R (1997) J Biol Chem 272:20283
143. Ilg EC, Schäfer BW, Heizmann CW (1996) Int J Cancer 68:325
144. Xia L, Stoll SW, Liebert M, Ethier SP, Carey T, Esclamado R, Carroll W, Johnson TM, Elder JT (1997) Cancer Res 57:3055
145. Glenney JR, Kindy MS, Zokas L (1989) J Cell Biol 108:569
146. Wicki R, Franz C, Scholl FA, Heizmann CW, Schäfer BW (1997) Cell Calcium 22:243
147. Song W, Zimmer DB (1996) Brain Res 721:204
148. Huber M, Siegenthaler G, Mirancea N, Marenholz I, Nizetic D, Breitkreutz D, Mischke D, Hohl D (2005) J Invest Dermatol 124:998
149. Segre JA, Bauer C, Fuchs E (1999) Nat Genet 22:356
150. Lee W, Mitchell P, Tjian R (1987) Cell 49:741
151. Manabe M, O'Guin WM (1994) Differentiation 58:65
152. Reddy SPM, Chuu Y-J, Lao PN, Donn J, Ann DK, Wu R (1995) J Biol Chem 270:26451
153. Jacob F (1977) Science 196:1161
154. Kartasova T, Darwiche N, Kohno Y, Koizumi H, Osada S-I, Huh N-H, Lichti U, Steinert PM, Kuroki T (1996) J Invest Dermatol 106:294
155. Gandarillas A, Watt FM (1995) Mamm Genome 6:680
156. Chang C-H, Scott GK, Kuo W-L, Xiong X, Suzdaltseva Y, Park JW, Sayre P, Erny K, Collins C, Gray JW, Benz CC (1997) Oncogene 14:1617
157. Oettgen P, Alani RM, Barcinski MA, Brown L, Akbarali Y, Boltax J, Kunsch C, Munger K, Libermann TA (1997) Mol Cell Biol 17:4419
158. Abraham JM, Wang S, Suzuki H, Jiang H-Y, Rosenblum-Vos LS, Yin J, Meltzer SJ (1996) Cell Growth Differ 7:855
159. Austin SJ, Fujimoto W, Marvin KW, Vollberg TM, Lorand L, Jetten AM (1996) J Biol Chem 271:3737
160. Steven AC, Steinert PM (1994) J Cell Sci 107:693
161. Jarnik M, Kartasova T, Steinert PM, Lichti U, Steven AC (1996) J Cell Sci 109:1381
162. Hohl D (1990) Dermatologica 180:201
163. Bucher P (1990) J Mol Biol 212:563
164. Mantovani R (1998) Nucleic Acids Res 26:1135

165. Fischer DF, Backendorf C (2005) Methods Mol Biol 289:303
166. Lallemand C, Bayat-Sarmadi M, Blanchard B, Tovey MG (1997) J Biol Chem 272:29801
167. Lallemand C, Palmieri M, Blanchard B, Meritet JF, Tovey MG (2002) EMBO Rep 3:153
168. Tvedt KE, Kopstad G, Haugen OA, Halgunset J (1987) Cancer Res 47:323
169. Chandra S, Fewtrell C, Millard PJ, Sandison DR, Webb WW, Morrison GH (1994) J Biol Chem 269:15186
170. Hardingham GE, Chawla S, Johnson CM, Bading H (1997) Nature 385:260
171. Dobi A, v Agoston D (1998) Proc Natl Acad Sci USA 95:5981
172. Corneliussen B, Holm M, Walterson Y, Onions J, Hallberg B, Thornell A, Grundström T (1994) Nature 368:760
173. Onions J, Hermann S, Grundström T (1997) J Biol Chem 272:23930
174. Chiang S-Y, Welch J, Rauscher FJ, Beerman TA (1994) Biochemistry 33:7033
175. Hennings H, Michael D, Cheng C, Steinert P, Holbrook K, Yuspa SH (1980) Cell 19:245
176. Watt FM, Matey DL, Garrod DR (1984) J Cell Biol 99:2211
177. De Robertis EM (1994) In: Duboule D (ed) Guidebook to the homeobox genes. Oxford University Press, Oxford, p 11
178. Maconochie M, Nonchev S, Morrison A, Krumlauf R (1996) Annu Rev Genet 30:529
179. Efstratiadis A, Posakony JW, Maniatis T, Lawn RM, O'Connell C, Spritz RA, DeRiel JK, Forget BG, Weissman SM, Slightom JL, Blechl AE, Smithies O, Baralle FE, Shoulders CC, Proudfoot NJ (1980) Cell 21:653
180. Tagle DA, Stanhope MJ, Siemieniak DR, Benson P, Goodman M, Slightom JL (1992) Genomics 13:741
181. Grosveld F, Blom van Assendelft G, Greaves D, Kolias G (1987) Cell 51:975
182. Kunkel TA (1985) Proc Natl Acad Sci USA 82:488
183. Purschke WG, Müller PK (1994) Biotechniques 16:264

Protein Binding Microarrays for the Characterization of DNA–Protein Interactions

Martha L. Bulyk

Division of Genetics, Department of Medicine, Department of Pathology,
Division of Health Sciences & Technology (HST),
Brigham & Women's Hospital and Harvard Medical School,
Harvard Medical School New Research Bldg., Room 466D, 77 Avenue Louis Pasteur,
Boston, MA 02115, USA
mlbulyk@receptor.med.harvard.edu

1	Introduction	66
2	Development of Protein Binding Microarrays	68
3	Proteins for Examination by Protein Binding Microarrays	70
4	Resources Required for Protein Binding Microarray Experiments	71
5	Design of Double-Stranded DNAs to Use in Protein Binding Microarray Experiments	72
6	Options in Immobilizing Double-Stranded DNAs to the Slide Surface	74
7	DNA Microarray Quality	75
7.1	DNA Purification and Printing Buffer	75
7.2	Microarray Data Quality Control	75
8	Determination of the DNA Binding Specificities of Proteins with Protein Binding Microarray Experiments	76
8.1	Protein Binding Microarray Experiments	76
8.2	Analysis of Protein Binding Microarray Data	77
8.2.1	Quantification of the Microarray Signal Intensities and Quality Control	77
8.2.2	Identification of the Significantly Bound Spots	78
8.2.3	Identification of the DNA Binding Site Motif from the Protein Binding Microarray Data	80
9	Prediction of Functional Roles of Transcription Factors from Protein Binding Microarray Data	80
9.1	Cross-Species Conservation of Protein Binding Microarray-Derived Transcription Factor Binding Sites	80
9.2	Functional Category Enrichment of Predicted Target Genes	81
9.3	Analysis of Publicly Available Gene Expression Datasets to Identify Conditions in Which a Significant Fraction of Protein Binding Microarray-Derived Target Genes are Differentially Expressed	81
10	Applications of Protein Binding Microarrays	82
11	Outlook	83
References		84

Abstract A number of important cellular processes, such as transcriptional regulation, recombination, replication, repair, and DNA modification, are performed by DNA binding proteins. Of particular interest are transcription factors (TFs) which, through their sequence-specific interactions with DNA binding sites, modulate gene expression in a manner required for normal cellular growth and differentiation, and also for response to environmental stimuli. Despite their importance, the DNA binding specificities of most DNA binding proteins still remain unknown, since prior technologies aimed at identifying DNA–protein interactions have been laborious, not highly scalable, or have required limiting biological reagents. Recently a new DNA microarray-based technology, termed protein binding microarrays (PBMs), has been developed that allows rapid, high-throughput characterization of the *in vitro* DNA binding site sequence specificities of TFs, other DNA binding proteins, or synthetic compounds. DNA binding site data from PBMs combined with gene annotation data, comparative sequence analysis, and gene expression profiling, can be used to predict what genes are regulated by a given TF, what the functions are of a given TF and its predicted target genes, and how that TF may fit into the cell's transcriptional regulatory network.

Keywords Protein binding microarray · DNA binding site motif · DNA–Protein interactions · DNA binding specificity

Abbreviations
ChIP   Chromatin immunoprecipitation
Dam   DNA adenine methyltransferase
dsDNA   Double-stranded DNA
GST   glutathione *S*-transferase
PBM   Protein binding microarray
TF   Transcription factor

1
Introduction

The interactions between transcription factors (TFs) and their DNA binding sites are an integral part of the regulatory networks within cells. These interactions control critical steps in development and responses to environmental stresses, and in humans their dysfunction can contribute to the progression of various diseases. Much progress has been made recently in the accumulation and analysis of mRNA transcript profiles and genome-wide location profiles [1,2]. However, there is still much to be understood about the transcriptional regulatory networks that govern these gene expression profiles.

One step along the way to developing a parallel methodology for characterizing the sequence specificity of DNA binding domains has been the use of 96-well plates for determining the "binding site signatures" of selected domains displayed on phage. Streptavidin-coated 96-well plates were bound by biotin-tagged sequences degenerate in two of three positions of a triplet binding site. Binding was measured in a semiquantitative manner using ELISA, and the resulting data for the 12 degenerate sequences were compiled to generate a binding site signature [3]. Another version of this methodology, employing luciferase fusion proteins, has been employed recently [4]. The primary limitation to such a binding site signature analysis is that one needs to start with a consensus or near-consensus sequence. In addition, a problem with simply compiling data from such partially degenerate sequences is that it assumes that all the base pairs of the DNA recognition site are acting in a completely independent fashion, when in reality there may be synergistic or destructive interference between different positions of a recognition site [5–8]. Therefore, the resulting binding site signatures may not accurately reflect the actual DNA binding specificity.

The development of DNA microarrays [9,10] has revolutionized mRNA expression analysis, and along with whole-genome sequencing of microbial and eukaryotic genomes has enabled various functional genomic technologies and systems-oriented analyses. Other array-based technologies include protein microarrays [11] for analysis of protein–protein interactions and interactions between proteins and small molecules, and microarrays of small molecules [12] for analysis of protein–ligand interactions.

DNA microarray-based readout of chromatin immunoprecipitation, also known as "ChIP-chip" or "genome-wide location analysis", is currently the most widely used method for identifying *in vivo* genomic binding sites for TFs in a high-throughput manner [13–16]. However, ChIP has some inherent caveats that can make the determination of a TF's DNA binding specificity difficult [17]. Indeed, some ChIP experiments do not result in significant enrichment of bound fragments in the immunoprecipitated (IPed) sample, and thus do not permit identification of the DNA sites bound *in vivo* [17,18]. Another recently developed method that takes advantage of DNA microarrays for the identification of *in vivo* binding sites of TFs utilizes tethered DNA adenine methyltransferase (Dam) [19]. This approach has been used to identify *in vivo* binding sites in *Drosophila* [20] and *Arabidopsis* [21]. However, it does not permit high-resolution mapping of binding sites, because methylation by the tethered Dam can extend over a few kilobases from the TF binding site [19].

Although *in vitro* selections have permitted the sampling of a large number of potential DNA binding sequences [22], the resulting sites provide only a partial view of the DNA binding specificity of the protein, as typically only the highest affinity binding sites are retained. It is possible that lower affinity DNA sites are functionally significant in transcriptional regulation of gene expression. For example, lower affinity sites may be responsible for the differences in function of two TFs that bind with high affinity to the same site (such as the *Drosophila* homeodomain proteins *even-skipped* and *fushi-tarazu*, or the murine homeodomain proteins Hmx1 and Nkx2.5) [23,24]. Although highly quantitative, surface plasmon resonance is not currently scalable to a large number of samples [25].

2
Development of Protein Binding Microarrays

Bulyk and colleagues have recently developed a new, highly parallel, *in vitro* microarray technology, termed protein binding microarrays (PBMs), for high-throughput characterization of the sequence specificities of DNA–protein interactions. In PBM experiments, a DNA binding protein of interest is expressed with an epitope tag. This tag serves a dual purpose: (1) it allows for purification of the expressed DNA binding protein, and (2) the epitope-tagged DNA binding protein is then applied to a dsDNA microarray. The protein-bound microarray is washed gently to remove any nonspecifically bound protein, and then stained with a primary antibody specific for the epitope tag (Fig. 1).

Shown in Fig. 2 is an example of a PBM in which a GST-tagged yeast TF was bound to a microarray printed with PCR products representing essentially all intergenic regions in the *Saccharomyces cerevisiae* yeast genome. Through PBM experiments using these whole-genome yeast intergenic microarrays, Bulyk and colleagues identified the DNA binding site sequence specificities of the yeast TFs Abf1, Rap1, and Mig1 (Fig. 3). For Abf1 and Rap1, DNA binding site motifs derived from the PBM data were highly similar to binding site motifs derived from ChIP-chip data [17]. Moreover, analysis of the Mig1 PBM data resulted in a match to the known binding site motif

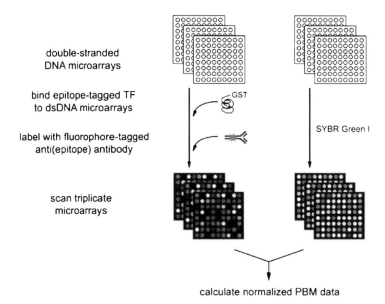

Fig. 1 Scheme of protein binding microarray experiments. (Reproduced from [26] with permission from Nature Publishing Group.)

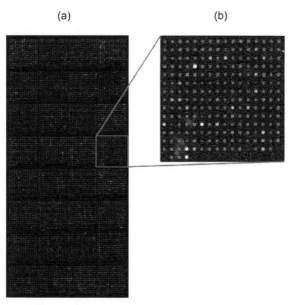

Fig. 2 Example of a PBM in which a GST-tagged yeast TF was bound to a whole-genome yeast intergenic microarray printed with PCR products. **a** Whole-genome yeast intergenic microarray bound by Rap1. The fluorescence intensities of the spots are shown in false color, with *white* indicating saturated signal intensity, *red* indicating high signal intensity, *green* indicating moderate signal intensity, and *blue* indicating low signal intensity. **b** Zoom-in on a portion of the whole-genome yeast intergenic microarray bound by Rap1. (Reproduced from [26] with permission from Nature Publishing Group.)

for Mig1 [26], while analysis of the ChIP-chip data [17] did not. In addition to previously identified targets, Abf1, Rap1, and Mig1 bound to numerous putative new target intergenic regions, many of which were upstream of previously uncharacterized open reading frames. Comparative sequence analysis indicated that many of these newly identified sites are highly conserved across five sequenced *sensu stricto* yeast species, and thus are likely to be functional *in vivo* binding sites that potentially are utilized in a condition-specific manner [26].

Importantly, the PBM technology allows the determination of the binding site specificities of known or predicted TFs in a single day, starting from the purified TF. Moreover, as with other microarray experiments, the PBM technology is highly scalable, allowing many PBM experiments to be performed in parallel. The PBM experiments themselves are neither time-intensive nor laborious; a single person can perform PBM experiments on a few TFs per day.

The PBM technology has several key advantages over high-throughput *in vitro* selection (a.k.a. SAGE-SELEX) methodology [27]. First, PBM data are more quantitative, since the signal within each spot on the microarray corresponds to numerous DNA–protein binding events. In addition, nonbinding sequences can be identified. Finally, PBMs can provide an extensive, if not complete, reference table of each DNA binding site sequence variant and its relative preference; the number of sequence variants examined is limited only by the number of features on the microarray.

3
Proteins for Examination by Protein Binding Microarrays

The Bulyk laboratory has successfully used TFs epitope-tagged with GST in PBM experiments using yeast intergenic microarrays (Figs. 2 and 3) [26], and also TFs expressed with the FLAG tag in PBM experiments using microarrays spotted with short synthetic dsDNAs [26]. The size of GST, combined with the use of a polyclonal Alexa488-conjugated anti-GST antibody, likely contributes to the high signal intensities achieved in those PBM experiments. Nevertheless, since GST can self-dimerize [28], other epitope tags may be preferable.

Another group has recently performed PBM experiments using the N-terminal domain of the *Drosophila* TF Extradenticle directly labeled with the fluorophore Cy3 at a unique cysteine [29]. Yet another group, using directly labeled TFs for binding to dsDNA microarrays, found that the TF Jun C-terminally labeled with Cy5-dC-puromycin was capable of interacting with

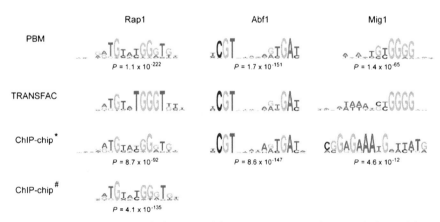

Fig. 3 DNA binding site motifs determined from PBMs compared to motifs derived from ChIP-chip data and from TRANSFAC. Sequence logos were generated essentially as described previously [52]. Group specificity scores were calculated as described in [48]. "*" indicates Rap1, Abf1, and Mig1 ChIP-chip data from Lee et al. [17], and "#" indicates Rap1 ChIP-chip data from Lieb et al. [13]. Although the Mig1 binding site motif derived from the ChIP-chip data has a statistically significant group specificity score, it is not a match to either the TRANSFAC or PBM Mig1 motif. (Reproduced from [26] with permission from Nature Publishing Group.)

its protein partner Fos, while Jun labeled at internal lysines did not bind to Fos [30]. Although direct labeling of the protein obviates the need for an antibody staining step, care must be taken to ensure that the incorporated fluorophore does not interfere with DNA binding or any protein–protein interactions necessary for DNA binding. Another possibility for labeling the protein would be to express it as a fusion to a fluorescent molecule, such as green fluorescent protein (GFP).

Overexpression of proteins in *Escherichia coli* is frequently performed, particularly when fairly large collections of proteins are being examined, because it is an inexpensive expression system that can produce high yields from relatively small cell culture volumes. Even though posttranslational modifications may be important for native protein function, many biochemical studies of TFs, and in particular their DNA binding specificities, are performed on proteins expressed in and purified from *E. coli*. However, certain proteins, particularly those larger than ∼80 kDa, may be difficult to overexpress in *E. coli* [31]. Alternatively, one could use an expression system that is biologically more similar to the organism whose DNA binding protein is being examined; for example, eukaryotic TFs could be expressed either *in vivo* in mammalian or insect cells, or *in vitro* in rabbit reticulocyte lysates. Finally, if a full-length protein is difficult to produce, one can attempt to increase the chances of successful expression and purification by instead expressing just its DNA binding domain plus any necessary protein–protein interaction domain(s).

4
Resources Required for Protein Binding Microarray Experiments

It is important to keep in mind the nature of the protein under examination; if a TF is being examined, then one needs to be sure to use dsDNA microarrays. Bulyk and colleagues have implemented PBMs on two different microarray platforms: robotically printed microarrays, and *in situ* synthesized oligonucleotide microarrays. Each of these two different platforms has its own unique advantages and disadvantages that include both technical and community accessibility issues, as discussed below.

The robotically printed dsDNA microarrays allow one to ensure that the material spotted onto the glass microarray slides is indeed double-stranded. Synthesis of oligonucleotides for use in PCRs or primer extension reactions can be performed or ordered by almost any laboratory. Although the cost of synthesis of a large set of oligonucleotides, and in some cases subsequent PCRs, can be great, the sequences present on the microarrays can be determined by the individual investigator and thus microarrays custom-designed for a particular research topic can be made fairly readily. This is accomplished by purifying the dsDNAs, after which there is sufficient material to print thousands of microarrays, which will reduce the long-term per-experiment costs for examining DNA binding proteins in the PBM experiments.

Moreover, most researchers have access to DNA microarraying facilities, if not at their own institution, then through another institution that provides microarraying services for a fee. For production of the whole-genome yeast intergenic DNA microarrays used in PBM experiments [26] as shown in Fig. 2, an OmniGrid® 100 microarrayer (Genomic Solutions, Ann Arbor, MI) equipped with Stealth 3 pins (Telechem International, Sunnyvale, CA) was used to spot DNA onto Corning® GAPS II or UltraGAPS 25 × 75-mm aminosilane-coated glass slides (Fisher Scientific). Approximately 0.7 nl DNA solution was deposited at each spot. Other slide types can potentially be used. Bulyk and colleagues have found that Corning® GAPS II and UltraGAPS slides result in low slide background in both PBM experiments and staining with SYBR Green I. Likewise, DNA microarray scanners are readily available in most departments or institutions. A ScanArray 5000 microarray scanner (Perkin Elmer, Boston, MA), which is equipped with a variety of laser and filter sets and permits microarrays to be scanned at a range of different laser power intensities or photomultiplier tube (PMT) gain settings, was used in PBM experiments [26] as shown in Fig. 2.

Nevertheless, not all users may have access to robotically printed dsDNA microarrays for use in PBM experiments. In addition, ultimately one might wish to have more features per microarray than typical microarraying robots can print onto standard 1 × 3-inch glass slides. Thus, Bulyk and colleagues have further developed the PBM technology using a commercially available microarray platform, available from Agilent Technologies, Inc., which already has the capacity to synthesize at least ~44 000 features per microarray. Note that Agilent microarrays are created by ink-jet synthesis as 60-mer single-stranded oligonucleotide microarrays [32], which subsequently need to be double-stranded (see next section) for use in PBMs to examine dsDNA binding proteins. Other platforms offer even higher densities; for example, Nimblegen uses micromirror arrays to synthesize microarrays of long oligonucleotides [33–35] at densities of up to ~760 000 features per microarray. Alternatively, in-house micromirror array synthesizers could be used to create high-density oligonucleotide microarrays. One group recently synthesized such self-hairpinning high-density oligonucleotide microarrays for use in PBM experiments [29].

5
Design of Double-Stranded DNAs to Use in Protein Binding Microarray Experiments

The key choice to be made in choosing what DNAs to print onto slides for use in PBM experiments is whether one wishes to synthesize a relatively low complexity microarray for directed experimentation on one or a small family of DNA binding proteins [36], or to synthesize a higher complexity microarray [26, 29, 37, 38] (Philippakis A, Qureshi A, Berger MF et al., personal communication) for examination of a broader set of proteins. In certain situations one might be able to restrict oneself to lower complexity microarrays that would be less expensive to produce if one were manufacturing the microarrays in-house, instead of having them synthesized by a commercial vendor.

Here I describe the design and synthesis of microarrays spotted with PCR products representing essentially all intergenic regions of the *S. cerevisiae* yeast genome, as the resulting microarrays can be used broadly [26], including for analysis of uncharacterized proteins, and have been described in multiple publications [13, 15, 17, 18]. These microarrays were printed with PCR products ~ 60–1500 bp long, covering essentially all noncoding regions of the *S. cerevisiae* yeast genome [15]. These whole-genome yeast intergenic microarrays were used in PBM experiments in order to identify the DNA binding site specificities of the *S. cerevisiae* TFs Rap1, Abf1, and Mig1 [26].

Microarrays spotted with coding regions are also expected to aid in identifying the sequence-specific binding properties of DNA binding proteins, despite the fact that it is currently thought that most *in vivo* regulatory sites will be located in non-protein-coding regions. Since PBM experiments are an *in vitro* technology, as long as there is sufficient sequence space represented on the DNA microarrays, one can expect to be able to derive a good approximation of the DNA binding site motif from the PBM data. Indeed, it is actually not necessary to utilize microarrays spotted with amplicons representing genomic regions from the same genome as the DNA binding protein of interest, but rather one can use microarrays spotted with a different genome's sequence. Nevertheless, one could use a genome-specific microarray, such as a promoter microarray [39] or a CpG island microarray [40], as long as such microarrays covered a sufficient amount of binding site sequence space.

The use of microarrays spotted with PCR products has the advantage of covering much sequence space with relatively few spots. However, inherent in those arrays are two key limitations. First, a single intergenic region may be bound once or multiple times at high, medium, or low affinity, depending upon the number and type(s) of candidate binding sites present within a given spotted intergenic region. Currently the measured fluorescence intensity of a spot cannot distinguish between these possibilities. Second, given the variation in probe lengths on the intergenic microarrays, a spot with a single binding site embedded in a long sequence will receive a less significant *P*-value than a spot with an identical binding site embedded in a shorter sequence [41].

Therefore, one may wish to consider instead using a microarray synthesized with short synthetic dsDNAs [26, 29, 36, 38]. Such dsDNAs can be made from single-stranded oligonucleotides either by primer extension [26, 36–

38, 41] or by self-hairpinning [29, 38]. Bulyk and colleagues have performed successful PBMs using microarrays spotted with synthetic dsDNAs ranging from ~35 to ~60 base pairs [26, 36, 38]. Another group has performed PBMs using microarrays synthesized *in situ* with hairpinned 34-mer oligomers containing a 14-bp double-stranded hairpin region [29].

Recently, "all k-mer"-style synthetic DNA microarrays have been described [29, 38] for use in PBMs, allowing the analysis of the binding profile for all k-mers up to $k = 8$ to 10 [29] or $k = 10$ to 12 [38] on a single 1×3-inch microarray. Such coverage of binding site sequence space can be accomplished by the synthesis of high-density oligonucleotide arrays [29, 37] or by a compact universal DNA design [38]. Briefly, with the use of high-density arrays, each individual k-mer can be situated on a distinct feature or spot on the array. However, since the number of possible k-mers can become very large for longer motifs, the number of such required spots can become greater than the number of spots that can be manufactured by robotic printing on a single 1×3-inch microarray [10, 32]. Therefore, instead of devoting a unique spot to each k-mer, one can instead employ a compact representation of k-mers [38]. In a compact universal design, for a given double-stranded DNA of length l significantly longer than the motif width k, each spot will contain $l - k + 1$ k-mers, when k-mers are considered in an overlapping fashion [38]. The key difference distinguishing the compact universal microarray technology over prior technologies is that all possible DNA sequence variants can be represented on DNA microarrays in a space- and cost- efficient manner, so that only a minimal number of individual DNA sequences and individual DNA spots need to be synthesized [38]. Importantly, "all k-mer"-style synthetic DNA microarrays, either those with each spot representing a single k-mer or those with a compact universal design, can be applied to the study of any proteins from any genome of interest.

6
Options in Immobilizing Double-Stranded DNAs to the Slide Surface

There are a few options for the immobilization of dsDNAs to the slide surface. Generally, the DNAs either can be attached randomly by UV cross-linking [42] or they can be end-attached, either by a reactive group at one of the DNA termini [36] or by *in situ* synthesis of arrays of oligonucleotides [9, 32, 33] that are subsequently double-stranded [37, 38]. In theory, end-attachment should allow the DNAs to not be kinked and to be maximally accessible for interaction with DNA binding proteins. However, gentle UV cross-linking can work well too (Bulyk and colleagues, unpublished results). Such a UV cross-linking protocol (i.e., millijoules setting) would need to optimize the two opposing issues of: (1) ensuring that the DNA structure is as unperturbed as possible, i.e., ideally most DNA molecules will have just one cross-link to the slide surface; and (2) ensuring that most spotted DNA molecules will be attached to the slides.

All three types of dsDNA immobilization have been used successfully to create microarrays used in PBM experiments [26, 36]. In the first method, the dsDNAs were end-attached to amine-reactive slides through the use of amino-tagged universal primers, as described previously [36]. In the second method, unmodified dsDNAs can be spotted onto various other types of slides, such as polylysine slides (Bulyk and colleagues, unpublished results) or GAPS II or UltraGAPS slides (Corning), and covalently attached to the slides via UV cross-linking in a Stratalinker (Stratagene) [41]. Finally, *in situ* synthesized oligonucleotide arrays can be biochemically double-stranded either by primer extension [36–38] or by self-hairpinning [29, 38].

7
DNA Microarray Quality

7.1
DNA Purification and Printing Buffer

In their published study using whole-genome yeast intergenic microarrays in PBM experiments to identify the DNA binding site specificities of the *S. cerevisiae* TFs Rap1, Abf1, and Mig1 [26], Bulyk and colleagues used microarrays printed with PCR products ~ 60–1500 bp in length, covering essentially all noncoding regions of the *S. cerevisiae* yeast genome [15]. Those genomic regions were amplified by PCR, and the completed PCR reactions were precipitated with ammonium acetate and isopropanol, washed with 70% ethanol, dried overnight, and resuspended in 3 × SSC printing buffer at a DNA concentration of 100–500 ng/µl. Alternatively, the PCR products may be filtered with purification plates, such as 96-well MultiScreen® PCR Filter Plates (Millipore, Billerica, MA). The extra filtration provided by the MultiScreen® plates increases the purity of the dsDNA. Other printing buffers or additives such as Sarkosyl or betaine may aid in increasing the spot uniformity and thus in improving the morphology of the printed spots. The use of different slide types can also result in different spot morphologies with given printing buffers; care should be taken to ensure that the chosen printing buffer is compatible with the chosen slide type.

7.2
Microarray Data Quality Control

Spot uniformity and good spot morphology allow more accurate quantification of spot signal intensities, and ultimately the degree of sequence-specific

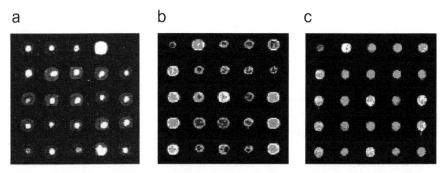

Fig. 4 Examples of DNA microarray spot quality. Identical portions of yeast intergenic microarrays printed onto Corning® GAPS II slides, processed in different ways (see below) before UV cross-linking, and then stained with SYBR Green I. Images have been false-colored as in Fig. 2. Examples of microarrays with poor spot quality are shown in (**a**) and (**b**). In both of these cases, the DNA is distributed nonuniformly, with either **a** high concentrations near the centers of spots, or **b** high concentrations along spot perimeters. Both of these microarrays resulted from two separate print runs, from which microarrays were UV cross-linked without first rehydrating and baking. An example of a good quality microarray is shown in (**c**). This microarray was rehydrated and then baked before being UV cross-linked. (Reproduced from [41] with permission from The Humana Press, Inc.)

binding of a given DNA binding protein to each spot. Severe problems with spot morphology frequently can be attributed to the choice of printing buffer and/or postprinting processing. Obviously problematic microarrays can be identified visually (Fig. 4) [41]. More subtle differences in spot quality can be identified through analysis of the quantified signal intensity data. Care should be taken to remove from consideration in subsequent data analysis steps any spots with too low DNA concentration to permit accurate quantification of the spot signal intensities, or spots in which the DNA is spread nonuniformly throughout the pixels. Various additional filtering criteria can be applied later during data analysis to remove from consideration any remaining spots that may be noisy even after removing spots with highly variable pixel signal intensities.

8
Determination of the DNA Binding Specificities of Proteins with Protein Binding Microarray Experiments

8.1
Protein Binding Microarray Experiments

Protocols for performing PBM experiments have been described in detail previously [26, 36, 38, 41]. Briefly, microarrays are first prewet and then blocked with a milk solution in order to minimize background. Milk can also be included in both the protein binding and antibody labeling reactions. Other blocking reagents may be suitable depending on the slide substrate on which the microarray was manufactured. Coverslips are typically used for the various microarray incubation steps. The use of LifterSlips™ coverslips helps to ensure a uniform distribution of the reaction mixture over the surface of the microarray. The microarrays are then incubated in a hydration chamber to prevent excessive evaporation of the reaction mixture under the coverslip.

The DNA binding protein of interest, typically at a final concentration in the range of approximately 20 nM, is initially preincubated with nonspecific DNA competitors. All incubations are typically performed for 1 h at room temperature [26, 41], but can be adjusted at the discretion of the user, as can the concentration of the protein in the binding mixture. Any necessary small molecules, such as zinc when examining zinc finger proteins, should be included in all binding and subsequent reactions and washes.

Once the preblocking and preincubation steps are completed, the microarrays are washed and then the protein binding reaction mixture is applied to the microarrays. During this time, fluorophore-conjugated antibody is preincubated in a milk solution. As with all fluorophores, all possible care should be taken to avoid photobleaching during the course of staining the microarrays. Alexa Fluor® 488 conjugated anti-glutathione S-transferase (anti-GST) polyclonal antibody (Molecular Probes) has been used successfully [26]. Other epitope tags and/or other antibodies conjugated with other fluorophores might also be used successfully. Once the protein binding step is completed, the microarrays are washed again, and then the preincubated antibody mixture is applied to the microarrays. Once the antibody staining step is completed, the microarrays are washed again, and then immediately spun dry in a table-top centrifuge. The dried microarrays are then ready for scanning using an appropriate laser and filter set (for Alexa Fluor™ 488, argon ion laser (488-nm excitation) and 522-nm emission filter).

8.2
Analysis of Protein Binding Microarray Data

8.2.1
Quantification of the Microarray Signal Intensities and Quality Control

In order to capture signal intensities for even very low signal intensity spots, while ensuring that subsaturation signal intensities are captured for as many spots as possible on the microarray, one can scan the microarrays at a number of different laser power (or PMT gain) settings, and then later integrate the data from these multiple scans [26, 36] using masliner software [43], as described below. The microarray TIF images can be quantified with microarray analysis software such as GenePix Pro (Axon Instruments, Inc.). After image quantification, one typically calculates the background-subtracted median intensities for use in subsequent analysis. One can then calculate the relative signal intensity data over the full series of scans taken at multiple laser power settings [26, 36]. To accomplish this task in a semiautomated fashion, one can use masliner (MicroArray Spot LINEar Regression) software, which combines the linear ranges of multiple scans from different scanner sensitivity settings onto an extended linear scale [43]. In their experiments using whole-genome yeast intergenic microarrays, Bulyk and colleagues observed that the final PBM and SYBR Green I stained microarrays frequently had post-masliner fluorescence intensities that spanned 5 to 6 orders of magnitude [26].

After masliner processing, any low-quality spots, such as those with dust flecks, should be removed from further consideration [26]. Next, the data from each of the replicate microarrays are normalized according to total signal intensity, and then within each individual microarray the data are normalized sector-wise, according to their local region on the slide. The data are then normalized again so that the mean spot intensity is the same over all the sectors. After these signal intensity normalizations, a number of additional quality control filtering criteria are applied, including the removal of spots with highly variable pixel signal intensities that could result in noisy PBM data, or spots that do not have highly reproducible data over the replicate microarrays. Additional *ad hoc* criteria (see [26] and [41]) can also further eliminate potentially noisy data points.

8.2.2
Identification of the Significantly Bound Spots

Once the PBM and SYBR Green I microarray data have been quantified, normalized, and filtered to remove noisy data points, the ratio of the mean PBM signal intensity divided by the mean SYBR Green I signal intensity can be used to identify the significantly bound spots [26]. Alternatively, one could use PBM data not normalized by the amount of DNA [29]. In general, a sequence-specific DNA binding protein is expected to bind preferentially to only a relatively small fraction of possible binding sites. Likewise, the remaining sequence variants are expected to be bound nonspecifically, as all DNA binding proteins are likely to exhibit some weaker affinity for nonspecific DNA binding sites [44]. One way to calculate the significance of binding, or P-value, for a given spot is to calculate its z-score [29, 37]. However, if more than a small percentage of spots are bound sequence-specifically, then another measure of significance, such as a pseudo-z-score [26, 41], may be more appropriate.

Details on how to calculate such pseudo-z-scores have been described previously [26, 41]. Briefly, the $\log_2$ of the ratios are LOWESS-normalized and then plotted as a histogram. The resulting distribution is expected to resemble a Gaussian distribution, corresponding to spots bound only nonspecifically, with specifically bound spots localizing to the upper tail of the distribution. The Gaussian-like distribution can then be used to calculate for each spot a pseudo-z-score that represents the probability that the spot belongs to the distribution of nonspecifically bound spots. Specifically, all values less than the mode of the Gaussian-like distribution are fit to a Gaussian function using the Mathematica software package (Wolfram Research, Inc., Champaign, IL). The pseudo-z-score for each spot is then calculated based on z, the number of standard deviations that the spot's log ratio departs from the mean of the Gaussian distribution [45].

Lastly, the pseudo-z-scores should be corrected for multiple hypothesis testing. Bulyk and colleagues previously employed the modified Bonferroni method [6, 46], using an initial $\alpha = 0.001$. Spots meeting or exceeding this significance threshold were considered significantly "bound" (Fig. 5a). Users may wish to consider spots at less stringent significance thresholds accordingly.

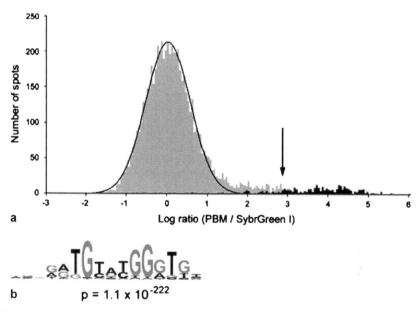

Fig. 5 Identification of the DNA binding site motif from the significantly bound spots. **a** Distribution of ratios of PBM data, normalized by SYBR Green I data, for the yeast TF Rap1 bound to yeast intergenic microarrays. The *arrow* indicates those spots passing a *P*-value cutoff of 0.001 after correction for multiple hypothesis testing. Indicated in *dark gray* are spots with an exact match to a sequence belonging to the PBM-derived binding site motif. **b** Sequence logo [52] of the PBM-derived motif for the yeast TF Rap1. (Reproduced from [26] with permission from Nature Publishing Group.)

8.2.3
Identification of the DNA Binding Site Motif from the Protein Binding Microarray Data

For the set of spots that are bound at the threshold significance level, one can then examine the corresponding set of DNA sequences for the likely DNA binding site motif of the given protein [26]. One might choose to search only the most significantly bound spots in order to minimize consideration of potentially false positive spots that would contribute noise to the motif-finding searches. For this set of input sequences, one typically uses a motif-finding algorithm, such as BioProspector [47], AlignACE [48, 49], MEME [50], or MDscan [51], in order to identify the DNA binding site motif of the protein. Since the binding site width of a TF is typically between 6 and 18 bp, the motif searches should be performed within this parameter range [26].

Once a motif has been identified by the given motif finder, one then needs to assess the likelihood of it being the DNA binding site motif of the given protein. This can be assessed statistically by calculating its group specificity score [48], which in this context indicates how specific the motif is to the set of bound spots as compared to all the spots on the microarray (for details on how to perform this calculation, and how to select the most likely TF binding site motif from the results, see [41]). In order to assess the statistical significance of the motifs resulting from this analysis, the results are compared against those from the analysis of a set of computational negative control sequence sets [26]. PBM-derived motifs with group specificity scores that are more significant than the group specificity scores of the corresponding computational negative control sets are considered to be good candidates for being the DNA binding site motif for the given DNA binding protein (Fig. 5b). Examples of the ranges of group specificity scores for computational negative controls and for actual PBM data for yeast TFs can be found in [26]. A graphical sequence logo [52] for each motif, such as those shown in Fig. 3, is often convenient for ease of visual examination of motifs and can be generated readily [53].

9
Prediction of Functional Roles of Transcription Factors from Protein Binding Microarray Data

9.1
Cross-Species Conservation of Protein Binding Microarray-Derived Transcription Factor Binding Sites

To find evidence supporting the hypothesis that the *S. cerevisiae* intergenic regions bound in the *in vitro* PBM experiments contain functional *in vivo* binding sites for the given TF, one can map the PBM-derived binding sites in *S. cerevisiae* to the orthologous positions in the sequence alignments of the *S. mikatae, S. kudriavzevii, S. bayanus*, and *S. paradoxus* genomes, which are the four other currently available sequenced yeast genomes of the yeast *sensu stricto* clade [54, 55]. Significant phylogenetic conservation suggests regulatory function of the PBM-derived TF binding sites.

9.2
Functional Category Enrichment of Predicted Target Genes

Analysis of a group of genes for enrichment for a particular functional annotation has been used previously to analyze sets of yeast genes that comprise particular gene expression clusters [56]. Each of the sets of intergenic regions bound in PBMs were examined to determine whether the groups of candidate target genes, located directly downstream of the bound intergenic regions, were overrepresented for particular functional groups of genes [48, 56]. The web-based tool FunSpec, with Bonferroni correction, was used for the statistical evaluation of these groups of genes, for groups of overrepresented gene and protein categories with respect to existing functional category information from a number of public and published databases [57]. FunSpec uses the hypergeometric distribution to calculate a *P*-value for functional category enrichment [48, 56]. Among the significantly enriched categories for the target genes derived from the Rap1 PBM data, many are consistent with the known regulatory functions of Rap1 [58], including the MIPS [59] functional classification categories for ribosome biogenesis ($p < 1.0 \times 10^{-14}$), protein synthesis ($p < 1.0 \times 10^{-14}$), structural constituents of the ribosome ($p < 1.0 \times 10^{-14}$), and cell growth and/or maintenance ($p = 3.5 \times 10^{-12}$).

9.3
Analysis of Publicly Available Gene Expression Datasets to Identify Conditions in Which a Significant Fraction of Protein Binding Microarray-Derived Target Genes are Differentially Expressed

Because different culture conditions often stimulate different cellular responses and coordinate changes in transcriptional regulation, the success of ChIP-chip experiments hinges on choosing those conditions in which the TF is expressed and actively regulating its target genes. PBMs, however, are free of this constraint and can identify TF binding site motifs and putative target genes irrespective of culture conditions. For example, Bulyk and colleagues used 643 publicly available *S. cerevisiae* gene expression datasets to identify conditions in which significant fractions of Abf1, Rap1, and Mig1 PBM target genes were differentially expressed. The conditions that exhibited the largest number of differentially regulated candidate target genes corresponded well with the known functions of each TF. For example, many Mig1 PBM target genes were downregulated at least 2.5-fold in glucose and fructose, compared to other carbon sources. These results show that together with expression profiling, PBM analysis can provide insight into the functions of particular TFs and identify conditions in which they are active *in vivo*. Therefore, analysis of the PBM-derived predicted target genes for conditions in which these genes are coregulated can further be used to suggest *in vivo* conditions for TF activity [26].

10
Applications of Protein Binding Microarrays

Two main types of studies have been performed using PBMs. In the first type, a family of closely related zinc finger proteins, including Zif268 (Egr1) and a number of artificial zinc finger proteins that arose from *in vitro* selections, were examined using a DNA microarray spotted with short synthetic dsDNAs specifically designed to interrogate all possible variants of a subset of the core binding site sequence (specifically, the central 3 bp of the Zif268 binding site) [36]. Because all the proteins were closely related, a DNA microarray could be designed to specifically examine the DNA binding site sequence variants expected to differ among the different proteins. A focused microarray, directed for a specific family of proteins, could be designed for other structural classes as well, as long as a consensus sequence or likely DNA binding site to use as a starting point for the family is known [60]. This approach can permit one to minimize microarray manufacturing costs by synthesizing only those dsDNAs thought to be most relevant for the family of proteins being examined. Similarly, if differences within a specific class of proteins are of interest, designing a focused microarray can permit one to thoroughly or near-thoroughly examine all likely binding site sequence variants of interest.

In the second type of study, a more generic DNA microarray was used to probe the sequence-specific binding of TFs representing a number of different structural classes of DNA binding proteins. The DNA microarrays were spotted with PCR amplicons representing essentially all intergenic regions of the *S. cerevisiae* yeast genome [26]. Instead of using phage display of DNA binding domains, that study used proteins expressed with an epitope tag. Such fusion proteins can be constructed readily using available genomic clone collections currently under construction for various model organisms as well as for the human genome. Because of the longer lengths of the spotted DNAs, the DNA binding site motifs of the query TFs were identified by motif-finding software [26, 41]. Since actual genomic sequences are represented on these arrays, one could also examine binding by multimeric protein complexes [30].

More recently, generic DNA microarrays spotted with short synthetic dsDNAs representing all *k*-mers have been described [29, 38] for use in PBMs, allowing the analysis of the binding profile for all k-mers up to $k = 8$ to 10 [29] or $k = 10$ to 12 [38] on a single 1×3-inch microarray. Such arrays have been used for the analysis of engineered polyamides [29] and for TFs [29, 38].

11
Outlook

There are predicted to be ~1850 TFs in the human genome [61], but only a very small fraction of them have well-characterized binding specificities. Likewise, most TFs from various model organisms are of as yet undetermined DNA binding specificities and in general their regulatory functions are not well understood on a genomic scale. The challenge will be to characterize their DNA binding specificities, so that their target genes and potential combinatorial modes of transcriptional regulatory control can be discovered. Continued improvements in the synthesis of high-density DNA microarrays will allow an even greater fraction of binding site sequence space to be surveyed.

In the future, PBM technology might potentially be used to derive DNA–protein binding affinities (K_d-values) for all possible DNA binding sites for a given TF. The affinities could either be interpolated from a set of reference DNAs, as has been done previously [36], or they could be determined from signal intensities from microarrays probed with a range of protein concentrations, as has recently been described for peptide interactions with protein microarrays [62]. Such binding data would be important for a better understanding of mechanisms of transcriptional regulation, such as potential competitive binding by TFs [63], and for improved prediction of *cis* regulatory elements in the genome [60].

Finally, in recent years, a number of efforts have been focused on attempting to predict TF binding sites using structural information on the protein or related DNA–protein complexes. Some of these studies have attempted to determine what "recognition rules" or "recognition code" may exist that stipulate what DNA base pairs are likely to be bound by what amino acids in the context of a particular structural class of DNA binding proteins. These approaches have come from analysis of databases of well-characterized DNA–protein interactions [64–68], computer modeling [69, 70], or experiments employing *in vitro* selection from a randomized library, either of the DNA base pairs or the amino acid residues implicated in sequence-specific binding [3, 71, 72]. However, there is no obvious, simple code like the genetic code, and any recognition rules that might exist are likely to be a quite degenerate "probabilistic code" [5] and highly dependent upon the docking arrangement of the protein with its DNA binding site [73]. Such efforts will be greatly aided by the further development of high-throughput technologies for identifying TF–DNA binding site interactions, so that much larger datasets can be generated for analyses required to decipher any degenerate probabilistic codes or to be used as training sets for developing improved DNA binding site prediction algorithms. Studies like these would allow us to understand better the biophysical determinants of observed DNA–protein interactions, and perhaps to glimpse the related selective pressures that underlie observed evolutionary changes in regulatory proteins and their target DNA binding sites.

Acknowledgements I thank Michael F. Berger and Tom Volkert for technical assistance. This work was supported in part by National Institutes of Health grants from the National Human Genome Research Institute to M.L.B. (R01 HG002966, R01 HG003420, and R01 HG003985).

References

1. Wyrick J, Young R (2002) Curr Opin Genet Dev 12:130
2. Lockhart DJ, Winzeler EA (2000) Nature 405:827
3. Choo Y, Klug A (1994) Proc Natl Acad Sci USA 91:11168
4. Hallikas O, Palin K, Sinjushina N et al (2006) Cell 124:47
5. Benos P, Bulyk M, Stormo G (2002) Nucleic Acids Res 30:4442
6. Bulyk M, Johnson P, Church G (2002) Nucleic Acids Res 30:1255
7. Lee M-L, Bulyk M, Whitmore G, Church G (2002) Biometrics 58:981
8. Man TK, Stormo GD (2001) Nucleic Acids Res 29:2471
9. Pease AC, Solas D, Sullivan EJ et al (1994) Proc Natl Acad Sci USA 91:5022
10. Schena M, Shalon D, Davis RW, Brown PO (1995) Science 270:467
11. MacBeath G, Schreiber SL (2000) Science 289:1760
12. MacBeath G, Koehler AN, Schreiber SL (1999) J Am Chem Soc 121:7967
13. Lieb JD, Liu X, Botstein D, Brown PO (2001) Nat Genet 28:327
14. Iyer VR, Horak CE, Scafe CS et al (2001) Nature 409:533
15. Ren B, Robert F, Wyrick JJ et al (2000) Science 290:2306
16. Reid JL, Iyer VR, Brown PO, Struhl K (2000) Mol Cell 6:1297
17. Lee T, Rinaldi N, Robert R et al (2002) Science 298:799
18. Harbison CT, Gordon DB, Lee TI et al (2004) Nature 431:99
19. van Steensel B, Henikoff S (2000) Nat Biotechnol 18:424
20. van Steensel B, Delrow J, Henikoff S (2001) Nat Genet 27:304
21. Tompa R, McCallum C, Delrow J et al (2002) Curr Biol 12:65
22. Oliphant A, Brandl C, Struhl K (1989) Mol Cell Biol 9:2944
23. Amendt B, Sutherland L, Russo A (1999) J Biol Chem 274:11635
24. Walter J, Dever C, Biggin M (1994) Genes Dev 8:1678
25. Udalova I, Mott R, Field D, Kwiatkowski D (2002) Proc Natl Acad Sci USA 99:8167
26. Mukherjee S, Berger MF, Jona G et al (2004) Nat Genet 36:1331
27. Roulet E, Busso S, Camargo AA et al (2002) Nat Biotechnol 20:831
28. Vargo MA, Nguyen L, Colman RF (2004) Biochemistry 43:3327
29. Warren CL, Kratochvil NC, Hauschild KE et al. (2006) Proc Natl Acad Sci USA 103:867–872
30. Doi N, Takashima H, Kinjo M et al (2002) Genome Res 12:487
31. Braun P, Hu Y, Shen B et al (2002) Proc Natl Acad Sci USA 99:2654
32. Hughes TR, Mao M, Jones AR et al (2001) Nat Biotechnol 19:342
33. Singh-Gasson S, Green RD, Yue Y et al (1999) Nat Biotechnol 17:974

34. Nuwaysir EF, Huang W, Albert TJ et al (2002) Genome Res 12:1749
35. Albert TJ, Norton J, Ott M et al (2003) Nucleic Acids Res 31:e35
36. Bulyk ML, Huang X, Choo Y, Church GM (2001) Proc Natl Acad Sci USA 98:7158
37. Bulyk ML, Gentalen E, Lockhart DJ, Church GM (1999) Nat Biotechnol 17:573
38. Berger MF, Philippakis AA, Qureshi AM, He FS, Estep PW 3rd, Bulyk ML (2006) Nature Biotechnol (in press)
39. Odom DT, Zizlsperger N, Gordon DB et al (2004) Science 303:1378
40. Weinmann AS, Yan PS, Oberley MJ et al (2002) Genes Dev 16:235
41. Berger MF, Bulyk ML (2006) In: Bina M (ed) Gene mapping, discovery, and expression (Methods Mol Biol). Humana, Totowa, New Jersey
42. DeRisi J, Penland L, Brown PO et al (1996) Nat Genet 14:457
43. Dudley A, Aach J, Steffen M, Church G (2002) Proc Natl Acad Sci USA 99:7554
44. Berg OG, Winter RB, von Hippel PH (1981) Biochemistry 20:6929
45. Taylor J (1997) An introduction to error analysis. University Science Books, Sausalito, CA
46. Sokal R, Rohlf R (1995) Biometry: the principles and practice of statistics in biological research. Freeman, New York
47. Liu X, Brutlag D, Liu J (2001) Pac Symp Biocomput 127
48. Hughes JD, Estep PW, Tavazoie S, Church GM (2000) J Mol Biol 296:1205
49. Roth FP, Hughes JD, Estep PW, Church GM (1998) Nat Biotechnol 16:939
50. Bailey T, Elkan C (1995) Proc Int Conf Intell Syst Mol Biol 3:21
51. Liu X, Brutlag D, Liu J (2002) Nat Biotechnol 20:835
52. Schneider TD, Stephens RM (1990) Nucleic Acids Res 18:6097
53. Crooks GE, Hon G, Chandonia JM, Brenner SE (2004) Genome Res 14:1188
54. Kellis M, Patterson N, Endrizzi M et al (2003) Nature 423:241
55. Cliften P, Sudarsanam P, Desikan A et al (2003) Science 301:71
56. Tavazoie S, Hughes J, Campbell M et al (1999) Nat Genet 22:281
57. Robinson M, Grigull J, Mohammad N, Hughes T (2002) BMC Bioinformatics 3:35
58. Planta RJ (1997) Yeast 13:1505
59. Mewes H, Frishman D, Guldener U et al (2002) Nucleic Acids Res 30:31
60. Michelson AM, Bulyk ML (2006) Mol Syst Biol 2:2006.0018
61. Venter JC, Adams MD, Myers EW et al (2001) Science 291:1304
62. Jones RB, Gordus A, Krall JA, MacBeath G (2006) Nature 439:168
63. Pierce M, Benjamin KR, Montano SP et al (2003) Mol Cell Biol 23:4814
64. Desjarlais JR, Berg JM (1992) Proc Natl Acad Sci USA 89:7345
65. Desjarlais JR, Berg JM (1992) Proteins 12:101
66. Jacobs G (1992) EMBO J 11:4507
67. Suzuki M, Yagi N (1994) Proc Natl Acad Sci USA 91:12357
68. Mandel-Gutfreund Y, Baron A, Margalit H (2001) Pac Symp Biocomput 139
69. Pomerantz JL, Sharp PA, Pabo CO (1995) Science 267:93
70. Pomerantz JL, Pabo CO, Sharp PA (1995) Proc Natl Acad Sci USA 92:9752
71. Choo Y, Klug A (1994) Proc Natl Acad Sci USA 91:11163
72. Rebar EJ, Pabo CO (1994) Science 263:671
73. Pabo CO, Nekludova L (2000) J Mol Biol 301:597

> # Accuracy and Reproducibility
of Protein–DNA Microarray Technology

Simon Field[1] · Irina Udalova[2] · Jiannis Ragoussis[1] (✉)

[1]Wellcome Trust Centre for Human Genetics, University of Oxford,
7 Roosevelt Drive, Oxford OX3 7BN, UK
Ioannis.ragoussis@well.ox.ac.uk

[2]Kennedy Institute of Rheumatology, Imperial College,
1 Aspenlea Road, London W6 8LH, UK

1	Introduction	88
1.1	Types of Microarrays Used for Protein–DNA Binding Studies	89
2	Preparation of Microarrays	92
2.1	Microarray Surface Chemistry	92
2.2	Sequence Selection	92
2.3	Length of DNA Sequence	93
2.4	Preparation of Duplexes for Spotting	93
2.5	DNA Printing	94
2.6	Slide Blocking	95
3	Protein Expression and Purification	95
3.1	Protein Expression Considerations	95
3.2	Purification Methods	96
4	Microarray Probing	97
4.1	Sensitivity	97
4.2	Equilibrium vs End-Point Affinity Reaction Conditions	99
4.3	Detection Methods	99
4.4	Wash Stringencies	101
4.5	Normalisation	102
5	Purified Protein or Cell Extracts?	103
5.1	Bacterial Cell Extracts	103
5.2	Mammalian Cell Extracts	104
6	Alternative Modes of Binding	107
7	Conclusions	107
	References	108

Abstract Microarray-based methods for understanding protein–DNA interactions have been developed in the last 6 years due to the need to introduce high-throughput technologies in this field. Protein–DNA microarrays utilise chips upon which a large number of DNA sequences may be printed or synthesised. Any DNA-binding protein may then be interrogated by applying either purified sample or cellular/nuclear extracts, subject to availability of a suitable detection system. Protein is simply added to the microarray slide surface, which is then washed and subjected to at least one further incubation with a labelled molecule which binds specifically to the protein of interest. The signal obtained is proportional to the level of DNA-binding protein bound to each DNA feature, enabling relative affinities to be calculated. Key factors for reproducible and accurate quantification of protein binding are: microarray surface chemistry; length of oligonucleotides; position of the binding site sequence; quality of the protein and antibodies; and hybridisation conditions.

Keywords Affinity · Cell extract · Consensus · DNA-binding domain · DNA–protein interactions · Hybridisation · Transcription factor · Protein purification · Protein-binding microarray

Abbreviations
Cy5	N,N'-Biscarboxypentyl-5,5'-disulphonatoindodicarbocyanine
DBD	DNA-binding domain
dIdC	Deoxyinosine–deoxycytosine
DNA	Deoxyribonucleic acid
DNA Pol I	DNA polymerase I, large (Klenow) fragment
EDTA	Ethylenediaminetetraacetic acid
EGTA	Ethylene glycol-bis(β-aminoethyl ether)-N,N,N',N'-tetraacetic acid
EMSA	Electrophoretic mobility shift assay
HEPES	4-(2-Hydroxyethyl)-1-piperazineethanesulphonic acid
HSV	Herpes simplex virus
IgG	Immunoglobulin G
IPTG	Isopropyl-β-D-thiogalactopyranoside
KCl	Potassium chloride
NaCl	Sodium chloride
NF-κB	Nuclear factor kappa B
Ni-NTA	Nickel-nitriloacetic acid
O.D.$_{600}$	Optical density at 600 nm
PAGE	Polyacrylamide gel electrophoresis
PBM	Protein-binding microarray
PBS	Phosphate-buffered saline
SDS	Sodium dodecyl sulphate
SSC	Saline sodium citrate
TF	Transcription factor

1
Introduction

High-throughput genomic technologies have enabled the sequencing of whole organisms [1, 2], the identification of genetic variation [3], the building of detailed transcript maps [4, 5] and exploration of the association between genetic variation and transcription [6]. The advent of systems biology [7] has contributed to the systematic annotation and ordering of gene networks and the integration of genomic, transcriptomic, proteomic and more recently metabolomic information.

Microarray technology is a major contributor to these achievements and is well established as the tool of choice for investigating patterns of gene expression in different tissue types or in cells subject to varying physiological conditions. The microarray-based experimental platform has now been extended to other genomic and proteomic applications, such as the identification of transcription factor (TF) binding regions in the genome (ChIP-on-chip method [8, 9]), as well as for identifying protein–protein interactions [10].

Protein–DNA microarrays have been developed to perform highly parallel investigations of protein–DNA interactions [11]. Transcription factors interact with DNA and are key elements in gene regulation, but quantitative surveys of TF-binding affinities to multiple sequence variants are sparse. For the majority of the 1962 known TFs [12], our knowledge of their DNA-binding specificities is limited to a few examples of DNA-binding motifs identified in non-systematic studies, and thus is often biased. The protein–DNA microarray is an ideal tool for systematic quantitative profiling of TF-binding specificities, essential for the identification of all relevant *cis*-regulatory elements in the genome and better understanding of gene regulation.

Protein–DNA microarrays utilise chips upon which a large number of DNA sequences may be printed or synthesised. To examine the DNA-binding specificities of any given TF, a sample of either purified protein (that may be tagged) or of a cellular protein extract is added to the slide surface. Binding to specific DNA motifs is then detected by incubation with labelled antibodies to the TF or its tag. The signal obtained is proportional to the level of protein bound to each DNA feature, enabling relative affinities to be calculated [13, 14]. In this review we will focus on key factors that determine the reproducible and accurate quantification of TF-binding specificities on protein–DNA microarrays.

1.1
Types of Microarrays Used for Protein–DNA Binding Studies

The first microarrays designed to interrogate protein–DNA interactions were presented in 1999 by Bulyk et al. [11]. In this paper the authors used single-strand (ss) arrays synthesised using light-directed in situ synthesis (3′ to 5′) by Affymetrix. A hexaethylene glycol (HEG) synthesis linker was used to link the DNA strands to the array. Single and double length linkers were tested. The ss sequence had a constant priming sequence at the 3′ end (close to the chip surface) that allowed a common primer to be annealed and used for on-chip primer extension reactions that produced double-stranded (ds) DNA molecules. The length of flanking sequences was between 5 and 20 bp. In this pioneering work the authors investigated DNA-binding proteins such as restriction enzymes and *dam* methylase.

Later, the same group used 37-bp oligonucleotides that were first extended using an amino-modified common primer and the resulting ds DNA was spotted on to slides (Gold Seal) [15]. The slides comprised activated, silanised glass. In this work the arrays (containing all variants of a 3-bp sequence) were used to identify binding sites for the zinc finger protein Zif268.

At the same time Mirzabekov's group used hydrogel slides to study the binding properties of the bacterial histone protein HU [16] and subsequently the Y-box protein p50 [17]. The group used hydrogel chips containing 4096 different ss and ds DNA molecules representing all possible hexamer sequences. The studies used directly labelled protein as well as melting curve analysis to identify moderate to weak binding sequences.

Activated, silanised slides were also used by Wang et al. [18]. They used a different approach to generate ds DNA molecules. A constant oligonucleotide containing a seven-base ss capture overhang at the 3' end and two reverse complementing sequences separated by a dT (amino-modified at C6) at the 5' end was pre-self-annealed and then annealed to target oligonucleotides. The target oligonucleotides (oligos) contained three parts, as follows, from 5' to 3': a 7-bp sequence complementary to the capture sequence of the constant oligo, a seven-base proximal flanking sequence followed by a 10-bp TF binding site, and a 7-bp distal flanking sequence. After the two oligos were annealed they were ligated. As a result a hairpin-like structure with an overhang was produced and this was spotted on activated silanised glass slides (Sigma). A final extension reaction using DNA polymerase I (Klenow) fragment was performed on the array to produce the final ds DNA molecules. The authors investigated the binding affinities of NF-κB p50 homodimers to wild-type and mutant sites.

We [13] used a strategy similar to the one used in [15], with the difference that the extended ds DNA molecules were spotted on Codelink (GE Healthcare) slides, and it was shown that on this particular type of surface the ds DNA is specifically covalently linked through the amino-modified base, thus increasing signal-to-noise ratios. The slides were used to perform binding analyses of NF-κB p50 and p52 as well as OCT-1 in combination with principal coordinate model analysis. Quantitative microarray binding data were validated using surface plasmon resonance and also correlated well with EMSA-derived affinities.

Egener et al. [19] used 25-bp ds DNA molecules generated by annealing of complementary oligos, which were sub-cloned into a vector, sequenced and amplified by PCR using amino-modified primers. The PCR products were subsequently spotted on Nexterion aldehyde slides (Schott). In this study the slides were used not only to determine sequence-specific binding of purified transcription factor AP-2, but also to detect the presence of AP-2 in breast cancer-derived MCF-7 cells. The limit of detection was determined to be in the 100–200 pg range (corresponding to about 2 fmol purified protein).

Mukherjee et al. combined the analysis of TFs (Abf1, Rap1 and Mig1) binding to ds genomic DNA microarrays with ChIP-on-chip data, revealing a degree of overlap but also differences between the two experiments [14]. It was clear that the data from protein-binding microarrays (PBMs)—as defined by Mukherjee et al.—provide valuable information, particularly for poorly understood TFs.

A different type of microarray-based assay was introduced by Fukumori et al. [20] and validated using the λ phage Cro repressor and p50. The assay involved stem loop DNAs formed by a 32-bp 5' region followed by a T_5 loop and a reverse complement 3' region. The stem loops were covalently bound to Codelink (GE Healthcare) slides using a T_{10} linker. The stem loop arrays were incubated with DNA-binding proteins and then subjected to Exo III digestion, which will digest the "bottom strand" of the stem loop from 3' to 5' up to the loop structure. If a protein is bound to the stem then these sequences will be protected and can be used to extend the strand back by Taq polymerase, whereby fluorescent dUTP can be incorporated and subsequently used to identify spots where the protein has bound. The advantage of this method is that it can be used in both homogeneous (solution) and microarray formats and does not require the use of antibodies against the protein.

Warren et al. [21] have generated in situ synthesised oligonucleotide arrays using a maskless array synthesizer (Nimblegen Systems, Madison). The oligos were self-complementing palindromes with a length of 14 bp and a TCCT loop. The palindromes contained two 3-bp flanks and an 8-bp binding site and were linked to the array surface by a T_{10} linker. The arrays contained 131 584 oligos and were used to determine the recognition profile of small DNA-binding molecules as well as a TF (Exd).

So far different technological platforms have been used to produce PBMs. It is obvious that in situ synthesis approaches are capable of producing higher density arrays compared to spotted arrays with feature numbers in the range of 10 000 to 1 000 000 or more (in the case of Affymetrix arrays). Such numbers will be required to interrogate a fully degenerate 10-mer (1 048 576 sequences). However, there are limitations in the length of oligonucleotides that can be synthesised in situ using the Affymetrix system at about 25 bp, while for example Nimblegen's technology allows the synthesis of oligos up to 85 bp long [22]. This is important, particularly when stem loop or similar approaches are applied to generate arrays able to interrogate 10–12-bp-long sites flanked by 10-bp or longer sequences. This type of limitation does not apply to spotted arrays, since the ds DNA molecules can be created in solution using primer extension or other techniques. The advantage of the latter approach is that a quality-controlled source of ds DNAs can be used to generate arrays that do not need further evaluation in that respect. In contrast, arrays based on solid-phase primer extension reactions will require additional quality control steps [11, 18].

2
Preparation of Microarrays

2.1
Microarray Surface Chemistry

Several types of slide and associated printing and scanning equipment are available commercially. The most frequently used surface chemistries are those that permit immobilisation of amino-modified DNA fragments: these include coatings such as carboxydextran matrix (via carboxyl groups) and *N*-hydroxysuccinimide esters (e.g. GE Healthcare's Codelink activated slides). The latter bind free amino groups at either end of suitably modified oligonucleotides, but not via other amine moieties elsewhere in the molecule. We have found that the polyacrylamide ester coating of the Codelink slides binds specifically to the amino group of modified oligonucleotides. In contrast, other coatings that are aminosilane, epoxy and aldehyde (amongst others) chemistry-based lead to unspecific binding of DNA molecules in parallel to forming specific covalent bonds with the amino-modified base. This is crucial in ensuring that DNA molecules are accessible to the protein(s) to be tested in their entire length and not bound to the glass surface in an unspecific way. As a result a very high degree of accuracy and reproducibility can be achieved, leading to a correlation coefficient of 93% between microarray and EMSA-derived binding affinities [13].

2.2
Sequence Selection

Lists of sequences to be tested experimentally can be generated by a greedy algorithm based on established consensus binding sites as for NF-κB or POU proteins, as implemented in [13] (http://neelix.molbiol.ox.ac.uk:8080/userweb/iudalova/cgi-bin/). The algorithm provides the minimum number of sequences for which data are required to make quantitative predictions of affinity in the entire sequence space, using a principal coordinates model to incorporate the effects of base-pair interactions rather than considering each position independently [23]. This is advantageous, not least because there are, for example, 524 800 unique 10-mers in one orientation alone.

Other sources of sequences can be simply drawn from published data sets, but it is important to perform further TF binding modelling using appropriate software packages such as the Sequence Alignment and Modelling System (SAM), http://www.cse.ucsc.edu/research/compbio/sam.html (for example used for the transcription factor AP-2 [19]).

Alternatively lists can be generated to represent all possible variations provided there is enough capacity on the array. In that respect, more recent improvements in printing technology that have increased slide capacity by reducing feature size, and lowered production costs by utilising in situ synthesis, mean that it may be feasible to generate microarrays with all DNA sequences of a particular length represented.

2.3
Length of DNA Sequence

We have used 34-bp fragments, which offer a compromise between ease of preparation (see below) and providing a length that minimises steric hindrance. Further to this, the protein binding site should be located as far as possible from the slide surface, although flanking sequences of the designed recognition site are also often important.

The design we have used incorporates a 20-bp spacer between the slide surface and the protein binding site, with a further 4-bp flanking sequence immediately following the binding site—see Sect. 2.4 below. Many DNA-binding proteins make non-specific contacts to the DNA phosphate backbone at positions in the flanking sequence around the recognition site [24, 25]; at least 4 bp should be provided. The flanking sequences around the protein binding site should be designed carefully to avoid creating alternative recognition sites. Final designs will be a compromise between sensitivity afforded by increasing sequence length, and reduced efficiency of primer extension across greater distances.

2.4
Preparation of Duplexes for Spotting

This is commonly achieved by annealing complementary sequences prior to printing on the microarray chip. The simplest method is to mix 100% complementary strands of equal length, one of which should be amino-modified, then denature and cool slowly to ensure complete annealing. A more economic approach is to use a shorter common modified sequence, to which longer oligonucleotides containing variable binding sites are annealed and extended to form fully double-stranded molecules Fig. 1.

A more flexible and cheaper alternative in the form of in situ synthesis on the slide surface is becoming more popular. Agilent has developed inkjet technology for the printing of custom sequences, providing a rapid means of generating arrays containing any sequence desired [26]. Creating ds DNA involves annealing a common oligonucleotide before performing a hybridisation on the slide using standard extension reagents, which may be monitored by measuring incorporation of a labelled base. In addition, the capacity of slides may be greater due to the potential to print features as small as 25 μm in diameter.

Common primer:
5'-H₂N-GGACCGATTGACTTGA-3'

+

Specific 34-mer:
3'-CCTGGCTAACTGAACTTCGANNNNNNNNNNTCGA-5'

where NNNNNNNNNN is a variant of GGRRNNYYCC (NFkB) or RYAKGNHAWY (POU) protein recognition sites.

↓ anneal

5'-H₃N-GGACCGATTGACTTGA-3'
     3'-CCTGGCTAACTGAACTTCGANNNNNNNNNNTCGA-5'

↓ extend with DNA Pol I (New England Biolabs)

5'-H₃N-GGACCGATTGACTTGA--> --> --> --> 3'
     3'-CCTGGCTAACTGAACTTCGANNNNNNNNNNTCGA-5'

↓ clean up, exchange into appropriate spotting buffer and print

¦-5'-H₃N-GGACCGATTGACTTGAAGCTNNNNNNNNNNAGCT-3'
     3'-CCTGGCTAACTGAACTTCGANNNNNNNNNNTCGA-5'

Fig. 1 Methodology for the production of duplexes from a single amino-tagged oligonucleotide (the common primer). Flanking sequences around the protein recognition site are *underlined*

2.5
DNA Printing

There are two general types of conventional array spotter: contact [27] and piezoelectric [28]. The spotting pins of the former make contact with the slide surface as material is deposited. A disadvantage is possible damage to the surface matrix during this process, together with scratches to which protein may bind non-specifically, which require stringent washes to remove. Piezoelectric instruments offer the potential for greater consistency in feature size and morphology, as the pins drop DNA without touching the slide surface. Instruments from different manufacturers may exhibit vastly different performance characteristics; guidelines for parameters such as DNA concentration, spot size and post-coupling conditions are platform-specific. We prepared DNA samples at a concentration of 20 μm for printing; the range 10–30 μm is suitable for Codelink slides (GE Healthcare). It may not be possible to detect protein if insufficient DNA is immobilised, and signal saturation could be a problem for some sequences if the DNA concentration is too high. High humidity (> 60%) during spotting is important for obtaining consistent feature morphology. We have found that 65% humidity and a distance between spots of 250 μm minimises the formation of "doughnut-shaped" spots and results in a large feature size without merging between adjacent spots. Analysis software is effective in dealing with variation in spot size across an array; see Sect. 4.5 for further information on normalisation for DNA concentration. It is common practice to incorporate at least four replicates of each sequence, together with a number of controls tailored to each protein to be tested, comprising a variety of known high- and low-affinity sequences if these are known.

2.6
Slide Blocking

Following printing and immediate post-spotting treatment based on empirically optimised protocols (often adapted from manufacturers' recommendations), the remaining reactive sites on slides should be chemically blocked. 0.1 M Tris/50 mM ethanolamine (no SDS) and 4× SSC/0.1% SDS are effective in blocking GE Healthcare Codelink activated slides. In addition to this, treatment with 2% Marvel™ milk protein in PBS is ideal for blocking further "sticky" sites on the surface.

3
Protein Expression and Purification

3.1
Protein Expression Considerations

It is important to consider a number of factors before producing protein to be interrogated on DNA-binding microarrays:

1. Is a full-length protein required, or are truncated DNA-binding domain(s) sufficient to obtain representative binding data?
2. Selection of expression system: higher yields from over-expression in bacteria versus potential toxicity and the ability to translate mammalian codons and fold the protein correctly, especially from full-length constructs.
3. How will the protein be purified? Expression tags are usually the simplest way to facilitate separation of the molecule of interest from endogenous proteins.
4. Co-purification of bound contaminating proteins may be an issue; even if there is no disruption to protein function, accurate yield determination is difficult if the sample contains impurities.

Many investigators design constructs based on the pET vector (Novagen), permitting expression of protein tagged with 6X histidine residues at one or both ends of the molecule. A number of commercially available kits enable rapid purification of the tagged protein via nickel-NTA (Novagen) chelation chromatography (see below). Several bacterial strains are suitable for protein over-expression; we have used inducible *Escherichia coli* BL21(DE3) cells (Novagen) which are simple to transform, produce high yields and are deficient in both *lon* and *ompT* proteases. Similar strains such as Rosetta (Novagen) may improve yield if there are a large number of mammalian codons which correspond to rarely expressed tRNAs in *E. coli*. The use of an inducible expression system minimises "leaky" expression which may be toxic to the host cell. We use a strain that features an additional plasmid encoding T7 lysozyme (BL21(DE3)pLysS; Novagen): this further represses basal expression of the target gene.

3.2
Purification Methods

Following successful transformation, a small starter culture should be grown. This should be diluted by a factor of 1/500 into either a single bulk volume of media (≥ 250 ml) or into several smaller vessels up to the required volume. We have found the latter to be most favourable for efficient expression of functional protein, with insignificant loss of plasmid through so-called bulk effects. An additional factor is that smaller bacterial pellets are generally easier to re-suspend. Details of induction conditions vary depending on host type and the nature of the protein, but typically bulk cultures of BL21 cells should be grown at 37 °C until O.D.$_{600}$ = 0.6–0.8. Expression is initiated with 0.2–1.0 mM IPTG, followed by growth for a further 2.5–5 h at 30 °C prior to harvesting.

Initial purification based on the (His)$_6$ expression tag is performed via Ni-NTA chromatography. However, a number of other bands may be visible when purified sample is analysed by SDS-PAGE, due to co-purification of proteins bound to the target or the presence of other histidine-rich molecules. The number and stringency of washes may help to eliminate a number of contaminating bands; we have found that 20 mM imidazole offers the best balance of purification level and yield of target protein, with elution by 125 mM imidazole. A proportion of the molecules present may be incorrectly folded or non-functional, but this cannot be ascertained by denaturing gel electrophoresis. We have optimised protocols for a second round of purification via DNA affinity chromatography, applicable to any DNA-binding protein with known high-affinity target sequences. A theoretical calculation of the number of DNA molecules required to bind a given amount of protein may not reflect the actual yields obtained, due to variable affinities and kinetics. Empirical optimisation is advised in order to produce reasonable yields with the minimum of reagents. To obtain comparable levels of protein for two different families of TFs, we have found that five times more DNA is re-

Accuracy and Reproducibility of Protein–DNA Microarray Technology 97

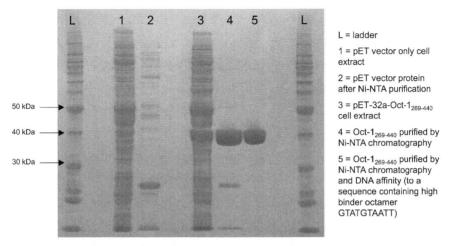

Fig. 2 SDS-PAGE analysis of proteins purified by two affinity chromatography methods

quired to isolate recombinant Oct-1 POU domains than for homodimers of human NF-κB DNA-binding domains (DBDs). DNA duplexes are prepared in a similar way to those required for spotting, but with 5′ biotinylation rather than amino modification. We anneal oligos of equal length (34 bases), which avoids the need to perform an extension reaction. These are immobilised on streptavidin-coated agarose (Sigma), which then forms the solid phase of a chromatography column. Binding is typically performed in 50 mM NaCl, with washes in 100–150 mM NaCl, followed by elution in 500 mM NaCl.

Purified samples should be exchanged into an appropriate buffer for DNA-binding studies, typically containing < 150 mM NaCl. The protein concentration may be established in a simple Bradford colorimetric assay, with standards of known concentration measured simultaneously. It can be seen that for the sample in lane 5 in Fig. 2, there is little contribution to the overall protein concentration from contaminating molecules.

4
Microarray Probing

There are a number of important considerations to be made when designing PBM experiments. These are detailed below.

4.1
Sensitivity

With little or no a priori knowledge of a given protein's binding characteristics, a degree of experimental optimisation is required. Protein concentration, reaction conditions and the nature of the sequences to be probed are variables that will have a significant effect on the relative affinities obtained. A problem is that high protein concentration and/or a long reaction period may be required to obtain a signal from the DNA sequences of lowest affinity. However, this may result in saturation of the signal from the highest binders, thus skewing binding ratios compared to experiments performed under different conditions. Lower concentrations obviate the latter phenomenon, but binding data may not be obtained for the lower affinity sequences if signal is not significantly higher than background (Fig. 3).

A few experiments are generally required to establish optimal binding conditions: three or four different protein concentrations are recommended to determine the approximate amount to use to achieve equilibrium binding for the majority of sequences. Initial binding data should be analysed by calculating ratios between concentrations; those giving linear increases/decreases in the average signal are suitable for use in subsequent experiments. The same applies for probing time.

We have used 1.5–5 µg (19–50 ng/µl) protein (monomer) per 80–100 µl reaction volume, probed for 1–2 h at room temperature in a humid chamber. The following buffer is recommended for initial experiments: 12 mM HEPES (pH 7.8), 80 mM KCl, 1 mM EDTA, 1 mM EGTA, 12% glycerol, 0.5 mg/ml poly(dIdC)–poly(dIdC), and 4% Marvel™ milk protein [29]. (HEPES is effective as a buffer at pH 6.8 to 8.2; the optimum for NF-κB protein binding is 7.5 to 8.0.) This involved application of protein directly onto the slide surface over a 26 × 20-mm area, with volume scaled up accordingly when required. At least 40 mM NaCl or KCl should be used. Experiments should be performed in a humid chamber.

If all or the vast majority of sequences appear to be positive, the stringency of protein–DNA binding could be increased by raising the salt concentration in the binding buffer. True sequence-specific binding at concentrations as high as 200 mM NaCl is possible, although 100–150 mM NaCl or KCl are more typical in PBM and EMSA experiments. This is another design consideration that should be optimised empirically.

B) Protein + DNA ⇌ Protein:DNA

C) Protein + DNA ⇌ Protein:DNA

A) Protein + DNA ⇌ Protein:DNA

Fig. 3 a Protein–DNA binding at equilibrium. **b** An excess of protein in the reaction may result in saturation of high binding sequences; the signal for very low-affinity sequences will be elevated relative to that of the high binders. **c** Binding to low-affinity sequences may not be detectable when the protein concentration is too low

4.2
Equilibrium vs End-Point Affinity Reaction Conditions

PBMs generally provide a "snapshot" of relative affinities at the reaction endpoint, although this varies from sequence to sequence as described above. A limitation of microarray technology is that measurements are not made in real time and do not provide kinetic data. As such, two sequences that appear to have similar affinities may in fact be very different in terms of association and dissociation rates. As long as the same reaction conditions are used, comparative affinity data remain valid. It may be possible to perform an experiment under approximately steady-state conditions, by carefully performing a number of experiments with different protein concentration and binding times, then selecting the conditions with the optimal balance of high sensitivity with minimal saturation of high-binding sequences. Analyses excluding the highest binders may be desirable to gain a clearer picture for the majority of sequences. Perfect conditions may be very difficult to achieve in reality. See below for more information on the effects of slide washing.

4.3
Detection Methods

In a perfect world it would be possible to label the protein of interest without disrupting its structure and function. However, this is seldom the case, especially when using truncated DBDs rather than full-length protein. Many dyes are bulky, meaning that even if successful site-specific labelling were performed, steric hindrance may continue to disrupt access and binding to DNA. PBM experiments therefore often involve multiple incubations with labelled secondary antibodies for the purposes of quantifying binding affinities. An advantage of the use of His-tagged proteins is the ready availability of anti-polyhistidine primary antibodies. We have found Santa Cruz Biotechnology, Inc.'s His-probe (H-15; raised in rabbit) to perform specifically in a 1 : 100 dilution, although the background signal after detection with labelled secondary antibody is higher than with an anti-NF-κB p52-specific IgG (Upstate). Antibody incubations were carried out in a humid chamber using 100 µl reaction volume containing 1X PBS and 2% Marvel™ milk protein, plus *either* 200 ng anti-polyhistidine (Santa Cruz) *or* 40 ng anti-NF-κB p52 (Upstate).

As our chosen anti-polyhistidine molecule is not available with a proven labelling system, we have performed experiments with a third hybridisation using a fluorophore-tagged anti-rabbit IgG. These were carried out in a humid chamber using 100 µl reaction volume containing 1X PBS and 2% milk protein (Marvel™), plus 1.5 µg Cy5-conjugated AffiniPure goat anti-rabbit IgG (H + L) (Jackson ImmunoResearch Laboratories, Inc.). The Cy5 label facilitates detection in most microarray scanners.

We have used the Axon Instruments GenePix 4000B slide scanner in conjunction with GenePix Pro 4.1 software. This instrument is compact and capable of scanning a chip in around 6 min. The sensitivity is not as great as that in other systems and only one slide may be placed in the instrument

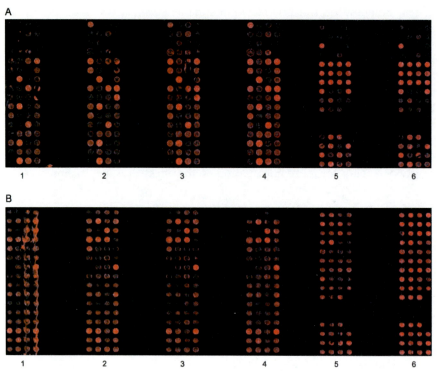

Fig. 4 a Part of a Codelink protein–DNA microarray containing six blocks of replicate spots (blocks 1–4 and 5–6 are replicates of two groups of DNA molecules) probed with the p52 DNA binding domain. Many features in block 1 have received less probing solution than blocks 2–4 during at least one of the incubations, due to incomplete surface coverage. In addition, it can be seen that spots in replicate blocks 5 and 6 are smaller than those in replicate blocks 1–4. This is due to variation in spotter pin performance, but has little effect because average pixel intensities across each spot are used in downstream data analysis. Some features are misshapen, which may be due to unintended contact with the slide surface during handling; analysis software identifies these as "bad" spots. **b** Codelink protein–DNA array (six replicate blocks as in **a**) probed with human Oct-1 POU domain. Poor mixing is responsible for low feature intensities in block 5; SYBR staining reveals comparable DNA concentrations between blocks 5 and 6. Block 1 (and the entire left side of the slide) appears to have been more affected by non-specific binding of protein and/or antibody than the rest of the slide. It may have received excess probing solution in one or more of the three incubations, or less wash buffer covered this region. Alternatively, a contaminant present on the slide surface may not have been removed during washing and subsequently cross-reacted with antibody

at a time. Slide layout/pattern (GAL) files must be written manually, but are simple to edit subsequently to incorporate changes in sequence positions.

This multiple-incubation system has provided consistent data for human Oct-1 POU and NF-κB p50 and p52 DBDs (up to 98% correlation between slides produced in the same printing run). However, variation in absolute values and ratios between different concentrations of the same protein may be observed for a number of reasons (Fig. 4):

- Differences in the concentration of functional protein; always use protein purified at the same time
- Incubation problems due to lack of consistent slide surface coverage or mixing
- Slides allowed to dry out between steps; try to avoid this
- Slight differences in washing times that skew relative affinities
- A specific "bad incubation" at control sequences that then influence overall data

The latter point highlights the importance of assessing variation between experimental replicates, most simply carried out by calculating the coefficient of variance (CV). Those that exceed an arbitrary threshold should be examined in greater detail, by checking each data point for obvious outliers and/or the array image. Some microarray analysis software may be capable of highlighting problem features automatically, usually requiring a degree of modification to quality control parameters due to hyper- or hypo-stringency.

4.4
Wash Stringencies

Washes after each incubation are of crucial importance. Clearly there is the need to remove protein bound non-specifically to features and the slide surface. There is also a requirement to wash away loosely associated protein, which will require different conditions depending on the nature of the DNA-binding protein of interest. There is a balance to be struck between removing too much protein from low-affinity sequences, whilst ensuring that the measured binding affinities will be an accurate reflection of true sequence-specific binding. We recommend 1% Tween in PBS buffer followed by 0.01% Triton-X100 in PBS for washing slides after initial probing with protein. The optimal number and length of washes should be determined empirically. As a general rule, lower stringency washes suffice for removal of primary and secondary antibodies applied in subsequent probing steps.

The following protocol is recommended, with each wash step requiring a minimum 50 ml volume and shaking for at least 3 min. Wash blocked slides with PBS/0.1% Tween-20 (1×) and PBS/0.01% Triton-X100 (1×). After TF binding, use PBS/1% Tween-20 (5×) and PBS/0.01% Triton-X100 (3×). Following primary antibody probing, wash with PBS/0.05% Tween-20 (3×) and

PBS/0.01% Triton-X100 (3×). Slides probed with labelled secondary antibody should also be washed with PBS/0.05% Tween-20 (3×) and PBS/0.01% Triton-X100 (3×), followed by PBS (1×) and a quick rinse in dH$_2$0 prior to drying and scanning.

The washing regime employed will have a large influence on whether the "snapshot" measurement reflects approximately steady-state or reaction endpoint conditions. For low-affinity sequences with rapid binding kinetics it is possible than no signal above background will be recorded. For sequences of higher affinity but also with fast dissociation rates, the data may prove to be an under-estimation of affinity at equilibrium. Care should be taken to rinse slides with PBS only prior to each hybridisation, as traces of viscous substances may affect mixing of binding buffer on the slide surface. Slides should be dried and scanned immediately after the final wash. For the Cy5 fluorophore, the emission maximum following laser excitation is at 635 nm (the red channel of most two-colour scanners).

4.5
Normalisation

There are several ways to ensure intra- and inter-chip comparisons between sequences remain valid, but first it is important to normalise for DNA concentration across the array. It is assumed that there will be relatively little variation in spot size and morphology between slides printed in the same run, meaning that it should only be necessary to assess relative DNA concentration across all of the features for only one slide per batch. We have used 1/2500 dilutions of 10 000X SYBR Green I or SYBR Gold (both Molecular Probes), a fluorescent dye that intercalates specifically between ds DNA (Fig. 5). Its emission spectrum maximum is at 532 nm (the green channel of most two-colour scanners). Note that the same laser power and/or PMT gain settings must always be used.

Raw data minus local background are provided by microarray analysis software. The median SYBR value should be taken and that feature assigned a value of 1. All other feature values should be adjusted by the same factor. Features within the same slide showing greater than twofold variation to the

Fig. 5 Image of a SYBR Gold-stained Codelink microarray scanned at 532 nm, showing four replicate blocks and variation in DNA concentration within them. The blocked slide was stained with 1/2500 SYBR Gold (Molecular Probes) in 2X SSC/0.1% Triton-X100 for 45 min at RT, and washed with 2X SSC/0.1% Triton-X100 (twice) and 2X SSC (twice)

median should be highlighted or excluded, as they will distort protein affinity data normalised on (divided by) these SYBR intensities. Unfortunately this system has severe limitations, as it assumes a linear relationship between DNA concentration and protein binding signal, whereas in reality the relationship is much more complex due to varying modes of binding, kinetics and relative affinities. There may be competition from neighbouring DNA molecules of higher affinity, for example, effectively quenching the signal obtained.

It should be noted that the SYBR staining system is also a hybridisation, subject to the same variations in mixing and surface coverage as any other probing procedure. Data points highlighted as "bad" by analysis software should be discarded; this applies to SYBR-staining and protein-binding experiments.

Once binding data have been normalised for DNA concentration, replicates should be averaged and checked for high CV values. Data that pass quality control are then normalised on a control sequence of medium affinity, identified from initial PBM experiments. We have assigned GGGGTTCCCC (NF-κB) and GTATGTAATT (POU) arbitrary values of 1000, with all other values changed by the same factor to achieve this. Sequences with relative affinity values significantly greater than 1000 are classified as high binders; those lower than 100 are non-binders. Medium-high binding sequences identified in this manner may then be employed in subsequent rounds of DNA affinity-based protein purification. Normalisation enables meaningful comparisons between technical replicates to be made. However, it is worth comparing raw values as well, as this will reveal whether there were experimental problems leading to significant differences in values between slides. Furthermore, raw data are required when calculating ratios between slides where different protein concentrations were used. Quality control alone must suffice for these analyses.

5
Purified Protein or Cell Extracts?

Purified protein is normally used for microarray experiments. On the other hand protein expressed in mammalian or other cell systems can also be used successfully. This can be important if there are concerns about protein conformation in bacteria, or system-specific post-translational modifications.

5.1
Bacterial Cell Extracts

We have found that it is possible to detect protein in bacterial cell extracts and nuclear extracts from mammalian cells. By using a specific antibody

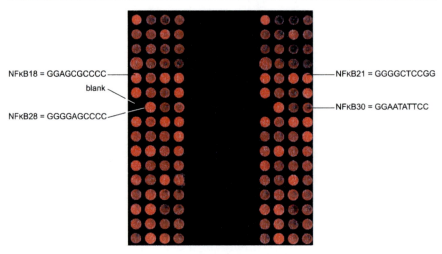

Fig. 6 Codelink protein–DNA microarray probed with bacterial cell extracts containing over-expressed human NF-κB p52 DBD detected with 40 ng anti-p52$_{4-19}$ (Upstate) and 1.5 µg Cy5-conjugated AffiniPure goat anti-rabbit IgG (H + L)(Jackson ImmunoResearch Laboratories, Inc.). BL21(DE3)pLysS cells were transformed with pET32a-p52$_{2-332}$, grown in 50 ml LB and induced with 200 mM IPTG at O.D.$_{600}$ = 0.6 before growth for a further 5 h at 30 °C. After harvesting, cells were resuspended in 1 ml buffer (50 mM Tris (pH 8.0); 0.1 mM EDTA; 50 mM NaCl; 10% glycerol; 0.01% NP-40), sonicated and cell debris pelleted. Around 0.8 ml cleared cell extract remained, of which 10 µl was used in this PBM experiment. NF-κB18 and NF-κB21 are classified as medium binders; NF-κB30 is a low binder

it may not be necessary to undertake time-consuming and costly protein purification procedures. Figure 6 illustrates part of a Codelink microarray probed with an extract obtained from BL21(DE3)pLysS cells (Novagen) transformed with pET32a-p52$_{2-332}$. The binding pattern closely resembles that shown in Fig. 7, albeit at reduced intensities. Data showed 92.7% correlation to that obtained with purified p52, but sensitivity was reduced (Fig. 9) due to a lower concentration of active p52 molecules. Correlations in excess of 98% have been recorded by using greater sample volumes in conjunction with modified protein expression induction conditions.

5.2
Mammalian Cell Extracts

Furthermore, we have similarly detected full-length p52 over-expressed in mammalian cells, using total protein extracts in conjunction with the detection system described above. Codelink protein–DNA microarrays were probed with total protein obtained from HEK 293 cells transfected with 4 µg eukaryotic expression vector encoding full-length human NF-κB p52

Accuracy and Reproducibility of Protein–DNA Microarray Technology 105

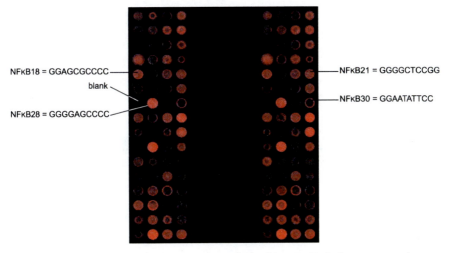

Fig. 7 Codelink protein–DNA microarray probed with p52. Each feature comprises an immobilised 34-bp DNA duplex containing a variant of κB or POU consensus protein binding sites at the end furthest from the slide surface. Purified human NF-κB p52 DBD (2.3 μg) was incubated on the array for 1 h. DNA-bound protein was detected with 40 ng anti-p52$_{4-19}$ (Upstate) and 1.5 μg Cy5-conjugated AffiniPure goat anti-rabbit IgG (H + L) (Jackson ImmunoResearch Laboratories, Inc.). Slides were washed with PBS/Tween-20 and PBS/Triton-X100 between incubations. Example κB binding site sequences are indicated. NF-κB18, NF-κB21 and NF-κB28 produced similar binding data and are classified as medium binders: the raw signal obtained from the latter was higher, but SYBR correction for DNA concentration produced similar normalised values. NF-κB30 is a low binder

(Fig. 8). Detection of DNA-bound protein was facilitated by incubation with 40 ng anti-p52$_{4-19}$ (Upstate) and 1.5 μg Cy5-conjugated AffiniPure goat anti-rabbit IgG (H + L) (Jackson ImmunoResearch Laboratories, Inc.). Correlation to binding data obtained with 2.3 μg purified NF-κB p52 DBD expressed in bacteria was up to 92.6%, indicating that relative rankings were maintained, but raw signal intensities were an average of 12-fold lower. Extracts derived from cells transfected with less than 1 μg plasmid contained insufficient protein for detection on the majority of sequences; a positive signal was obtained for only three sequences in the example of cells transfected with 125 ng plasmid. We have demonstrated that functional protein may be produced in both bacterial and eukaryotic systems, and that complex purification procedures are not a prerequisite for successful protein–DNA microarray experiments. This offers great potential for use in comparative studies of samples derived from resting and stimulated cells, in terms of protein levels. In addition, the use of antibodies specific to modified versions of the protein should permit an assessment of relative levels of phosphorylated and/or adenylated forms in cells subject to different physiological conditions.

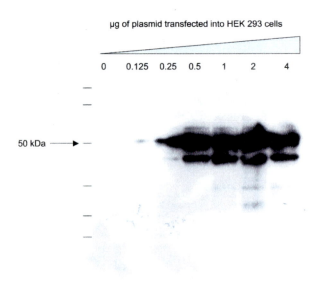

Fig. 8 Western blot analysis of total protein obtained from mammalian HEK 293 cells transfected with an expression plasmid encoding full-length human NF-κB p52. Recombinant protein was detected with anti-p52$_{4-19}$ (Upstate)

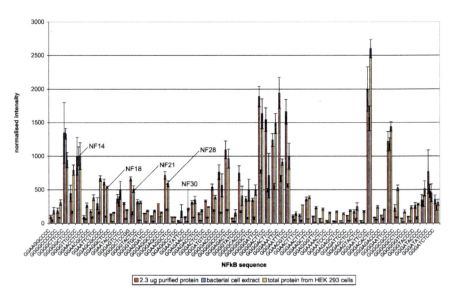

Fig. 9 Normalised binding data obtained for purified human p52 DBD, bacterial cell extracts containing p52 DBD and mammalian cell extracts containing over-expressed full-length p52 on Codelink protein–DNA microarrays. The sequence GGGGTTCCCC (NF-κB14) was assigned a value of 1000 in each experiment. NF-κB18, NF-κB21 and NF-κB28 are classified as medium binders (Fig. 7); NF-κB30 is a low binder

6
Alternative Modes of Binding

It should be noted that many TFs may not bind exclusively to one type of consensus sequence; alternative modes of sequence recognition have been described for Oct-1, for example. The first issue to consider is whether the full length or DBD of the protein is capable of binding to other sites on the designed DNA probes. Flanking sequences around the intended protein binding site should be designed in such a way that whole or partial recognition sites are not formed in isolation or by overlap with the binding site. If cell extracts are used, it may be possible that endogenous macromolecules associate with or covalently modify the protein of interest, causing changes to sequence selection preferences. Many DNA-binding proteins function as homo-multimers with more than one subset of target sequences. Purified protein may dimerise in solution or by a cooperative process induced by DNA binding.

A summary of Oct-1 recognition site types as an example is given in Table 1. It can be seen that extreme care must be taken in experimental design to avoid creating additional sites.

Table 1 Oct-1 binds at least five distinct types of recognition sequence. Bases in red indicate those contacted by a single POU domain (only one possible mode of binding shown for OCTA+). MORE and PORE sites are targets of dimeric Oct-1

Sequence subset	Example
Octamer	ATGCAAAT
OCTA+ TAATGARAT (from HSV ICP0 promoter)	CGAGCATGCTAATGATATTCTTC
OCTA- TAATGARAT (from HSV ICP4 promoter)	GAGGGCGGTAATGAGATACGA
MORE (More palindromic Oct factor recognition element)	ATGCATATGCAT
PORE (Palindromic Oct factor recognition element)	ATTTGAAATGCAAAT

7
Conclusions

Protein–DNA microarrays have developed into powerful, high-throughput tools for protein–DNA binding studies. They are able to contribute to the identification of TF binding sites, as well as the further quantitative profiling of TF DNA-binding specificities.

It is clear that arrays with the capacity to contain one million or more sequences will be required to perform a systematic profiling analysis. This is particularly important for transcription factors that exhibit various degrees of flexibility in binding site recognition, even if they belong to the same family, e.g. POU domain TFs [30, 31].

The information obtained from protein–DNA microarrays can be combined with data generated by alternative high-throughput approaches such as bacterial one-hybrid systems [32], variants of SELEX [33] or computational approaches such as enhancer element locator (EEL) [34]. The use of combinations of computational and experimental approaches [35, 36] can help to understand how proteins interact with DNA and allow the encoded information to be utilised in a way that leads to the development and maintenance of a fully functional organism.

Acknowledgements This work was funded by the Wellcome Trust.

References

1. Lander ES, Linton LM, Birren B, Nusbaum C, Zody MC, Baldwin J, Devon K, Dewar K, Doyle M, FitzHugh W, Funke R, Gage D, Harris K, Heaford A, Howland J, Kann L, Lehoczky J, LeVine R, McEwan P, McKernan K, Meldrim J, Mesirov JP, Miranda C, Morris W, Naylor J, Raymond C, Rosetti M, Santos R, Sheridan A, Sougnez C, Stange-Thomann N, Stojanovic N, Subramanian A, Wyman D, Rogers J, Sulston J, Ainscough R, Beck S, Bentley D, Burton J, Clee C, Carter N, Coulson A, Deadman R, Deloukas P, Dunham A, Dunham I, Durbin R, French L, Grafham D, Gregory S, Hubbard T, Humphray S, Hunt A, Jones M, Lloyd C, McMurray A, Matthews L, Mercer S, Milne S, Mullikin JC, Mungall A, Plumb R, Ross M, Shownkeen R, Sims S, Waterston RH, Wilson RK, Hillier LW, McPherson JD, Marra MA, Mardis ER, Fulton LA, Chinwalla AT, Pepin KH, Gish WR, Chissoe SL, Wendl MC, Delehaunty KD, Miner TL, Delehaunty A, Kramer JB, Cook LL, Fulton RS, Johnson DL, Minx PJ, Clifton SW, Hawkins T, Branscomb E, Predki P, Richardson P, Wenning S, Slezak T, Doggett N, Cheng JF, Olsen A, Lucas S, Elkin C, Uberbacher E, Frazier M et al (2001) Nature 409:860
2. Waterston RH, Lindblad-Toh K, Birney E, Rogers J, Abril JF, Agarwal P, Agarwala R, Ainscough R, Alexandersson M, An P, Antonarakis SE, Attwood J, Baertsch R, Bailey J, Barlow K, Beck S, Berry E, Birren B, Bloom T, Bork P, Botcherby M, Bray N, Brent MR, Brown DG, Brown SD, Bult C, Burton J, Butler J, Campbell RD, Carninci P, Cawley S, Chiaromonte F, Chinwalla AT, Church DM, Clamp M, Clee C, Collins FS, Cook LL, Copley RR, Coulson A, Couronne O, Cuff J, Curwen V, Cutts T, Daly M, David R, Davies J, Delehaunty KD, Deri J, Dermitzakis ET, Dewey C, Dickens NJ, Diekhans M, Dodge S, Dubchak I, Dunn DM, Eddy SR, Elnitski L, Emes RD, Eswara P, Eyras E, Felsenfeld A, Fewell GA, Flicek P, Foley K, Frankel WN, Fulton LA, Fulton RS, Furey TS, Gage D, Gibbs RA, Glusman G, Gnerre S, Goldman N, Goodstadt L, Grafham D, Graves TA, Green ED, Gregory S, Guigo R, Guyer M, Hardison RC, Haussler D, Hayashizaki Y, Hillier LW, Hinrichs A, Hlavina W, Holzer T, Hsu F, Hua A, Hubbard T, Hunt A, Jackson I, Jaffe DB, Johnson LS, Jones M, Jones TA, Joy A, Kamal M, Karlsson EK et al (2002) Nature 420:520

3. Sachidanandam R, Weissman D, Schmidt SC, Kakol JM, Stein LD, Marth G, Sherry S, Mullikin JC, Mortimore BJ, Willey DL, Hunt SE, Cole CG, Coggill PC, Rice CM, Ning Z, Rogers J, Bentley DR, Kwok PY, Mardis ER, Yeh RT, Schultz B, Cook L, Davenport R, Dante M, Fulton L, Hillier L, Waterston RH, McPherson JD, Gilman B, Schaffner S, Van Etten WJ, Reich D, Higgins J, Daly MJ, Blumenstiel B, Baldwin J, Stange-Thomann N, Zody MC, Linton L, Lander ES, Attshuler D (2001) Nature 409:928
4. Shoemaker DD, Schadt EE, Armour CD, He YD, Garrett-Engele P, McDonagh PD, Loerch PM, Leonardson A, Lum PY, Cavet G, Wu LF, Altschuler SJ, Edwards S, King J, Tsang JS, Schimmack G, Schelter JM, Koch J, Ziman M, Marton MJ, Li B, Cundiff P, Ward T, Castle J, Krolewski M, Meyer MR, Mao M, Burchard J, Kidd MJ, Dai H, Phillips JW, Linsley PS, Stoughton R, Scherer S, Boguski MS (2001) Nature 409:922
5. Johnson JM, Castle J, Garrett-Engele P, Kan Z, Loerch PM, Armour CD, Santos R, Schadt EE, Stoughton R, Shoemaker DD (2003) Science 302:2141
6. Stranger BE, Forrest MS, Clark AG, Minichiello MJ, Deutsch S, Lyle R, Hunt S, Kahl B, Antonarakis SE, Tavare S, Deloukas P, Dermitzakis ET (2005) PLoS Genet 1:e78
7. Oltvai ZN, Barabasi AL (2002) Science 298:763
8. Ren B, Robert F, Wyrick JJ, Aparicio O, Jennings EG, Simon I, Zeitlinger J, Schreiber J, Hannett N, Kanin E, Volkert TL, Wilson CJ, Bell SP, Young RA (2000) Science 290:2306
9. Martone R, Euskirchen G, Bertone P, Hartman S, Royce TE, Luscombe NM, Rinn JL, Nelson FK, Miller P, Gerstein M, Weissman S, Snyder M (2003) Proc Natl Acad Sci USA 100:12247
10. Zhu H, Snyder M (2003) Curr Opin Chem Biol 7:55
11. Bulyk ML, Gentalen E, Lockhart DJ, Church GM (1999) Nat Biotechnol 17:573
12. Messina DN, Glasscock J, Gish W, Lovett M (2004) Genome Res 14:2041
13. Linnell J, Mott R, Field S, Kwiatkowski DP, Ragoussis J, Udalova IA (2004) Nucleic Acids Res 32:e44
14. Mukherjee S, Berger MF, Jona G, Wang XS, Muzzey D, Snyder M, Young RA, Bulyk ML (2004) Nat Genet 36:1331
15. Bulyk ML, Huang X, Choo Y, Church GM (2001) Proc Natl Acad Sci USA 98:7158
16. Krylov AS, Zasedateleva OA, Prokopenko DV, Rouviere-Yaniv J, Mirzabekov AD (2001) Nucleic Acids Res 29:2654
17. Zasedateleva OA, Krylov AS, Prokopenko DV, Skabkin MA, Ovchinnikov LP, Kolchinsky A, Mirzabekov AD (2002) J Mol Biol 324:73
18. Wang JK, Li TX, Bai YF, Lu ZH (2003) Anal Biochem 316:192
19. Egener T, Roulet E, Zehnder M, Bucher P, Mermod N (2005) Nucleic Acids Res 33:e79
20. Fukumori T, Miyachi H, Yokoyama K (2005) J Biochem (Tokyo) 138:473
21. Warren CL, Kratochvil NC, Hauschild KE, Foister S, Brezinski ML, Dervan PB, Phillips GN Jr, Ansari AZ (2006) Proc Natl Acad Sci USA 103:867
22. Albert TJ, Norton J, Ott M, Richmond T, Nuwaysir K, Nuwaysir EF, Stengele KP, Green RD (2003) Nucleic Acids Res 31:e35
23. Udalova IA, Mott R, Field D, Kwiatkowski D (2002) Proc Natl Acad Sci USA 99:8167
24. Cho Y, Gorina S, Jeffrey PD, Pavletich NP (1994) Science 265:346
25. Michalopoulos I, Hay RT (1999) Nucleic Acids Res 27:503
26. Hughes TR, Mao M, Jones AR, Burchard J, Marton MJ, Shannon KW, Lefkowitz SM, Ziman M, Schelter JM, Meyer MR, Kobayashi S, Davis C, Dai H, He YD, Stephaniants SB, Cavet G, Walker WL, West A, Coffey E, Shoemaker DD, Stoughton R, Blanchard AP, Friend SH, Linsley PS (2001) Nat Biotechnol 19:342
27. Schena M, Shalon D, Davis RW, Brown PO (1995) Science 270:467
28. Okamoto T, Suzuki T, Yamamoto N (2000) Nat Biotechnol 18:438
29. Schreiber E, Matthias P, Muller MM, Schaffner W (1989) Nucleic Acids Res 17:6419

30. Li P, He X, Gerrero MR, Mok M, Aggarwal A, Rosenfeld MG (1993) Genes Dev 7:2483
31. Cleary MA, Herr W (1995) Mol Cell Biol 15:2090
32. Meng X, Brodsky MH, Wolfe SA (2005) Nat Biotechnol 23:988
33. Roulet E, Busso S, Camargo AA, Simpson AJ, Mermod N, Bucher P (2002) Nat Biotechnol 20:831
34. Hallikas O, Palin K, Sinjushina N, Rautiainen R, Partanen J, Ukkonen E, Taipale J (2006) Cell 124:47
35. ML (2003) Genome Biol 5:201
36. Michelson AM, Bulyk ML (2006) Mol Syst Biol 2:2006 0018

Identification and Characterization of DNA-Binding Proteins by Mass Spectrometry

Eckhard Nordhoff (✉) · Hans Lehrach

Department Lehrach, Max Planck Institute for Molecular Genetics, Ihnestrasse 73, 14195 Berlin, Germany
nordhoff@molgen.mpg.de

1	Introduction	113
2	Ionization, Instrumentation, and Ion Fragmentation	114
3	Sample Preparation	119
4	**Mass Spectrometry Based Methods**	123
4.1	Protein Identification in Sequence Databases	124
4.2	Protein Sequencing	128
4.3	Protein Modification Analysis	129
4.4	Analysis of Noncovalent Complexes	133
4.5	Protein Footprinting, Cross-Linking, and Surface Labeling	135
4.5.1	Protein Footprinting	136
4.5.2	Cross-Linking	138
4.5.3	Surface Labeling	141
4.6	Hydrogen/Deuterium Exchange	141
4.7	Combining MS with Liquid Chromatography and Electrophoresis	145
4.8	Combining MS with Affinity Purification Techniques	151
4.9	Combining MS with Biomolecular Interaction Analysis	157
4.10	Protein Quantification and Stable Isotope Labeling	158
5	Applications and Examples	169
5.1	Identification and Quantification	169
5.2	Posttranslational Modifications	174
5.3	Higher-Order Structures and Interactions	179
	References	183

Abstract Mass spectrometry is the most sensitive and specific analytical technique available for protein identification and quantification. Over the past 10 years, by the use of mass spectrometric techniques hundreds of previously unknown proteins have been identified as DNA-binding proteins that are involved in the regulation of gene expression, replication, or DNA repair. Beyond this task, the applications of mass spectrometry cover all aspects from sequence and modification analysis to protein structure, dynamics, and interactions. In particular, two new, complementary ionization techniques have made this possible: matrix-assisted laser desorption/ionization and electrospray ionization. Their combination with different mass-over-charge analyzers and ion fragmentation techniques, as well as specific enzymatic or chemical reactions and other analytical techniques, has led to the development of a broad repertoire of mass spectrometric methods that are now available for the identification and detailed characterization of DNA-binding proteins. These techniques, how they work, what their requirements and limitations are, and selected examples that document their performance are described and discussed in this chapter.

Keywords DNA-binding proteins · ESI MS · MALDI MS · Mass spectrometry · Stable isotope labeling

Abbreviations
2-DE	Two-dimensional gel electrophoresis
BIA	Biomolecular interaction analysis
cAMP	Cyclic adenosine monophosphate
CDIT	Culture-derived isotope tags
CE	Capillary electrophoresis
CHCA	α-Cyano-4-hydroxycinnamic acid
CID	Collision-induced dissociation
DHB	2,5-Dihydroxybenzoic acid
Dnmt2	DNA methyltransferase-2
ECD	Electron capture dissociation
ESI	Electrospray ionization
ETD	Electron transfer dissociation
HX	Hydrogen/deuterium exchange
ICAT	Isotope-coded affinity tag
ICPL	Isotope-coded protein tag
ICR	Ion cyclotron resonance
IRMPD	Infrared multiple photon dissociation
iTRAQ	Isobaric tag for relative and absolute quantification
LC	Liquid chromatography
m/z	Mass-over-charge
MALDI	Matrix-assisted laser desorption/ionization
MudPIT	Multidimensional protein identification technology
PMF	Peptide mass fingerprinting
PSD	Post-source decay
PTM	Posttranslational modification
RNAi	RNA interference
SA	Sinapic acid
SDS-PAGE	Sodium dodecyl sulfate–polyacrylamide gel electrophoresis
SELDI	Surface-enhanced laser desorption/ionization
SILAC	Stable isotope labeling by amino acids in cell culture
SLE	Systemic lupus erythematosus
SPR	Surface plasmon resonance
TAP	Tandem affinity purification
TEV	Tobacco etch virus
TOF	Time-of-flight

1
Introduction

The term DNA-binding proteins often refers only to proteins that bind to DNA strongly and for a long time. In this chapter, however, that definition is extended to also cover all proteins that bind to DNA for a short time, e.g., it includes all DNA-processing enzymes. Today, mass spectrometry (MS) is the most sensitive and specific technique available for protein identification and detailed characterization. Its applications cover all aspects from sequence and modification analysis to structure, dynamics, interactions, and relative and absolute quantification. Consequently, MS has become an important platform technology for the study of DNA-binding proteins. A prominent example supporting this notion is the identification of the nucleolus subproteome comprising at least 692 different proteins. In fact, just in the past 2 years, many hundreds of hitherto unknown proteins were identified by the use of MS as DNA-binding proteins that are involved in the regulation of gene expression, replication, or DNA repair. Another example that has gained considerable attention among molecular biologists is the proof that DNA methyltransferase-2 (Dnmt2), whose name and function are based on close sequence homology to authentic DNA cytosine methyltransferases, in fact is not a DNA-processing enzyme. With the aid of MS it could be shown that Dnmt2 in vivo, instead of DNA, methylates cytosine 38 in the anticodon loop of aspartic acid transfer RNA.

Over the past 10 years, MS has also changed our view on chromatin structure and regulation. A picture of a few known posttranslational modifications (PTMs) of the histone protein family, with little knowledge about their functions, has been turned into a combinatorial explosion of a still growing number of possible modifications and modification sites that are known or expected to be correlated with chromatin structure, function, and activity. Soon after this change was realized, a histone code was postulated that consists of series of site-specific modifications, which encode biological information. It is assumed that, if it exists, this code is part of the epigenetic information linked to proteins that can be inherited by following generations in addition to the genetic code. At the time this book is written, MS is used to answer the question of whether this code exists or not, and if it exists, how complex it really is.

In particular, two new, complementary ionization techniques have made the above success stories possible: matrix-assisted laser desorption/ionization (MALDI) and electrospray ionization (ESI). Their combination with different mass-over-charge (m/z) analyzers, ion fragmentation techniques, as well as specific enzymatic or chemical reactions and other analytical techniques, has led to the development of a broad range of MS methods for the identification and further characterization of proteins in general, and DNA-binding proteins in particular. To aid understanding of how these methods work, what their requirements are, and how they are applied, this contribution starts with a short description of ESI and MALDI, and the kind of MS instrumentation and ion fragmentation techniques that they are combined with. This is followed by a short description of the different MS sample preparation techniques available for the analysis of peptides and proteins, and what their requirements are. A description and discussion of the different MS-based methods, mentioned above, occupies the central part of this chapter. Thereafter, selected applications and examples document what is possible, and the problems and limitations encountered when using MS to identify DNA-binding proteins and explore their structures and functions.

2
Ionization, Instrumentation, and Ion Fragmentation

For analytical chemists, MS has been the first choice for the detection and characterization of all kind of molecules up to a molecular mass of approximately 1000 Da for many years. With MS, one can detect, distinguish, and unambiguously identify tiny amounts of closely related compounds in complex samples in a short time using fully automated workflows. Besides accurate molecular masses and relative ion abundances, tandem MS (or MS/MS) also provides structural information on molecular ions, which forms the basis for unambiguous compound identification.

To determine the mass of molecules by MS, they must first be ionized and transferred into the gas phase such that individual molecular ions are available for subsequent measurements of their mass-to-charge ratio (m/z). For many years, however, large molecules like proteins were not accessible to MS because the ionization techniques available resulted in their decomposition before the analysis was complete. This limitation was overcome in the late 1980s by the introduction of ESI [1, 2] and MALDI [3–5] (Fig. 1). Both techniques turned out to be very successful and their requirements and performances complementary [6–11]. As a result, in combination with different mass analyzers they soon became commercially available and together started a new era of protein analytics.

MALDI is a pulsed ion source that generates with each laser shot of a few nanoseconds pulse width a discrete ion package that contains predominantly singly charged (peptides), doubly charged (small and medium-size proteins), or triply charged (large proteins) analyte ions [4, 5, 12]. In contrast, ESI generates a continuous ion beam of multiply charged molecular ions (ca. one charge/1000 Da) [1, 2]. Consequently, MALDI was first coupled to a time-of-flight (TOF) analyzer [4], ideal for pulsed ionization events (Fig. 1a,b), and ESI to a quadrupole or triple-quadrupole mass spectrometer [1], which expects a continuous ion flow (Fig. 1c), or electrostatic ion traps that collect and analyze fractions thereof [13] (Fig. 1e).

Analysis of DNA-Binding Proteins by Mass Spectrometry 115

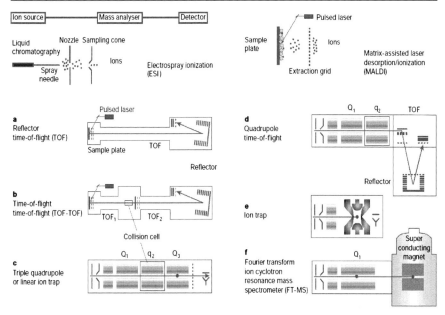

Fig. 1 Mass spectrometers used in proteome research. The *left* and *right upper panels* illustrate the ionization and sample introduction process in electrospray ionization (ESI) and matrix-assisted laser desorption/ionization (MALDI). **a–f** Different m/z analyzer configurations used in combination with ESI and/or MALDI. **a** In reflector time-of-flight (TOF) instruments, all ions are accelerated to high kinetic energy (e.g., 25 keV for singly charged ions) and are then separated along a defined flight path as a consequence of their different velocities ($E_{kin} = 1/2\,mV^2$). The ions are turned around in a reflector, which compensates for slight differences in kinetic energy resulting from the ion formation process, before they are detected. **b** The TOF–TOF instrument incorporates a collision cell between two TOF sections. Ions of a specific m/z are isolated in the first TOF section, selected, and then fragmented in the collision cell. The resulting fragment ions are analyzed by the following TOF-MS. **c** Quadrupole mass spectrometers scan ions by time-varying electric fields between four rods, which permit at a time a stable trajectory only for ions of a particular m/z. For MS/MS experiments, ions of a particular m/z are selected in a first section (Q1), fragmented in a collision cell (q2), and the fragments separated in Q3. In the linear ion trap, ions are captured in a quadrupole section, depicted by the *red dot* in Q3. They are then excited via resonant electric fields and the fragments are scanned out, creating the tandem mass spectrum. **d** The quadrupole TOF instrument combines the front part of a triple quadrupole instrument with a reflector TOF section for m/z analysis. **e** The (three-dimensional) ion trap captures the ions as in the case of the linear ion trap, fragments ions of a particular m/z, and then scans out the fragments to generate the tandem mass spectrum. **f** The FT-MS instrument also traps the ions, but does so with the help of strong magnetic fields. The figure shows the combination of FT-MS with the linear ion trap. The latter is used for efficient isolation and fragmentation and the FT-MS part for accurate m/z analysis. Reproduced from [7] with permission from © Nature Publishing Group, 2006

In the following years, the latter category of m/z analyzers has been combined with both ionization techniques [14], as have been FT-ICR MS (Fig. 1f), TOF MS in orthogonal configuration, and a series of hybrid instruments that combine a quadrupole MS or electrostatic ion trap with an orthogonal TOF [15–18] (Fig. 1d) or FT-ICR analyzer [19–23]. The most recent innovation has been a new electrostatic ion trap design, called orbitrap [24, 27], which in combination with an upfront linear ion trap [25, 26] provides excellent MS/MS and MS/MS/MS performance, impressive resolving power (> 50 000 FWHM), and a mass accuracy of 0.5–2 ppm for peptides that comes close to what, so far, has only been possible with very expensive FT-ICR MS instrumentation [20, 27–29]. This means that for a 1-kDa peptide the expected experimental error is only 0.0005–0.002 Da or 1–4 times the mass of an electron.

Using a high repetition rate laser combined with a high (atmospheric) or medium pressure source, a pulsed MALDI ion source can be used to generate a "quasi ion beam" [15], which can be scanned by quadrupole MS. On the other hand, in orthogonal configuration, the ion flux generated by ESI can be analyzed by TOF MS (pulsed ion extraction). These developments are a consequence of the complementary nature of the two ionization techniques which, ideally, should both be available for all current m/z analyzer techniques and any new ones being developed [11].

Of the latter category, ion mobility MS is an especially interesting candidate because it introduces an additional means of separation (two-dimensional separation). To achieve this, the molecular ions generated by ESI or MALDI migrate first through a low or medium pressure zone of defined length, which separates them according to their cross section, before they enter one of the established m/z analyzers, preferably TOF MS in the orthogonal configuration [30–33]. Ion mobility is an interesting analytical parameter for the separation of peptide and protein molecular ions because their cross section directly depends on their 3D structure, which makes it possible to separate different 3D structures of one and the same primary structure (identical m/z) inside the MS instrument, e.g., a stretched versus a globular compact conformation of a protein [32, 34]. For the analysis of complex peptide and protein samples, fractionation by ion mobility before separation by m/z promises a simple and efficient means to extend the resolving power [31, 34, 35].

MALDI-TOF MS is by far the fastest and also the most sensitive of all MS techniques available for the analysis of peptides and small proteins, due to the very short analysis times (fractions of a millisecond) and high ion transmission of TOF MS. In addition, it is not limited to the analysis of peptides, as are all currently available combinations of MALDI with FT-ICR MS, electrostatic ion traps, or quadrupole analyzers due to their limited m/z range. This limitation does not apply to ESI, which in contrast to MALDI delivers large molecules exclusively as highly charged molecular ions.

A major difference between ESI and MALDI is that ESI requires liquid samples whereas MALDI, so far, works best with crystalline samples. Another important difference is that MALDI requires a special matrix compound for the analysis of biological macromolecules, e.g., α-cyano-4-hydroxycinnamic acid (CHCA) for small and medium-size peptides, sinapic acid (SA) for large peptides and proteins, 2,5-dihydroxybenzoic acid (DHB) for glycosylated peptides and proteins, and 3-hydroxypicolinic acid for nucleic acids. It is noteworthy that all these compounds have been found solely empirically and that only a very few out of many hundred candidates turned out to be successful.

An important advantage of MALDI versus ESI MS is that with the former the data acquisition can be interrupted at any time and unconsumed sample recovered for later analyses. Furthermore, MALDI is more tolerant to sample contaminations such as salts, detergents, and chaotropic or reducing reagents [36]. However, there are also disadvantages when compared to ESI. In general, MALDI samples are less homogeneous leading to fluctuating analyte ion signals during data acquisition. In contrast, ESI has inherently higher sample homogeneity and analyte ion flux stability due to the sample being in the liquid state. These characteristics allow easier interfacing to ion transfer sensitive instruments such as FT-ICR mass spectrometers and potentially quantification solely based on ion signal intensities or peak area integrals [37]. In addition, liquid chromatography (LC) can be coupled online to ESI MS enabling fully automated workflows (Fig. 1).

As a consequence, for accurate molecular mass analysis of intact proteins, ESI in combination with an orthogonal TOF, electrostatic ion trap or FT-ICR MS is the preferred instrumentation if the samples are very clean and not too complex. If this is not the case or sample throughput is more important than mass accuracy, MALDI-TOF MS is the first choice. Another important advantage of ESI is its superior performance for the analysis of noncovalent complexes including DNA–protein interactions.

For the mass spectrometric analysis of peptides, the two ionization techniques are similarly complementary as they are for proteins [38]. Today, most mass spectrometers equipped with an ESI or a MALDI source provide the possibility of isolating analyte ions on the basis of their m/z ratio and, by different activation methods, transfer energy to them which results in their decomposition. Mass analysis of the resulting fragment ions can provide detailed structural information and is termed tandem-MS or MS/MS analysis (Fig. 2). Electrostatic ion traps and FT-ICR analyzers also provide multiple stages of ion isolation and fragmentation experiments (MS^n), of which MS^3 is especially useful for protein identification and modification analyses.

In MALDI-TOF MS, analyte fragmentation can be induced simply by increasing the laser irradiance several percent above the threshold used for MS analysis [39]. Under these conditions, the resulting fragmentation process is typically delayed such that most of the fragment ions are formed outside the

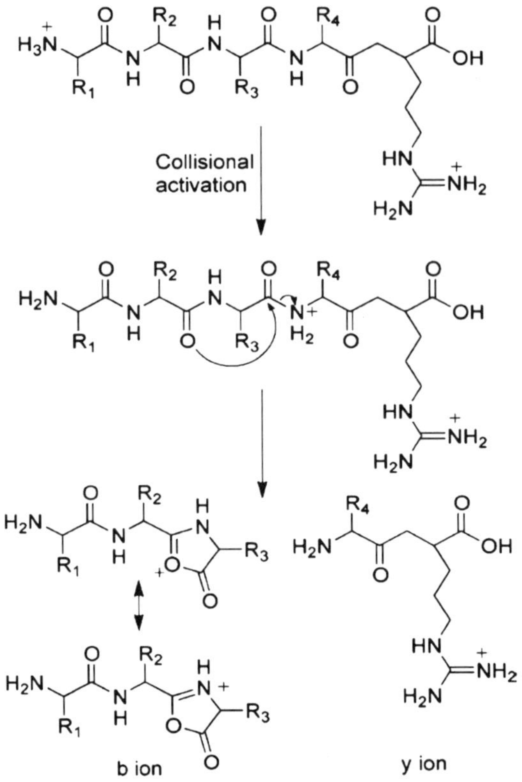

Fig. 2 Fragmentation of a doubly charged tryptic peptide resulting in b and y ions. Relocation of the N-terminal proton to promote charge-directed cleavage via collisional activation may occur to either N or O backbone atoms. Reproduced from [10] with permission from © Birkhäuser Verlag, 2006

ion source in the field-free drift region. Accordingly, these events are termed post-source decay (PSD). This fragmentation mechanism or high-energy collision-induced dissociation (CID) are used by MALDI-TOF MS/MS [40–42]. These two combinations are currently the most sensitive for MS/MS analyses of peptides. The resolving power and mass accuracy, as well as the control over the fragmentation reactions is, however, significantly lower compared to those of ESI or MALDI combined with hybrid orthogonal TOF, electrostatic ion traps, or FT-ICR MS.

The main advantage of using ESI instead of MALDI is the production of abundant doubly and triply charged peptide molecular ions, which are a lot easier to fragment in a controlled fashion, e.g., by low-energy CID [18, 43] or infrared multiple photon dissociation (IRMPD) [44, 45], than their singly charged relatives predominantly produced by MALDI. In addition, the two most powerful fragmentation techniques currently available, i.e., electron capture dissociation (ECD) [46, 47] and electron transfer dissociation (ETD) [48–52], are restricted to multiply charged precursor ions, and are therefore exclusively combined with ESI. With these techniques, an electron is transferred to or abstracted from the isolated molecular ions resulting in the formation of radicals, which are unstable and quickly fall apart [53–58]. Because the analyte ions lose one charge, MS/MS analyses are not possible for singly charged precursor ions. With regard to the information content of peptide fragment ion spectra, CID, IRMPD and ECD or ETD are complementary techniques, which for structure analyses are best combined, especially CID with ECD or ETD [48, 50].

Today, with MALDI and ESI MS, mid attomole to low femtomole amounts of peptides can be analyzed in MS and MS/MS mode. For proteins, however, especially if larger than 20 kDa, a minimum of 50–100 femtomole is required and MS/MS analyses are more difficult to conduct and far from any widespread routine application. This explains why most of the MS-based protein research is still done on the level of peptides, although this might change in the not too distant future. When studying DNA-binding proteins, however, the situation is somewhat different as one important class, the histones, is easy to isolate in large quantities and small enough to take advantage of the additional information that direct MS and MS/MS offers [51, 59–62].

3
Sample Preparation

This section summarizes the most important sample preparation techniques that have been developed for MALDI and ESI MS of peptides and proteins. For the study of DNA-binding proteins, this typically covers only the last steps of a long chain of sample preparation steps, which are necessary to isolate them out of their biological environment and provide them in a form that is suitable for MALDI or ESI MS. As the sample preparation of peptides and proteins for ESI or MALDI, with some exceptions, is little specific for their ability to bind to DNA or not, this section is not restricted to that feature.

Both ionization techniques, MALDI and ESI, deliver best results if the samples are pure and concentrated. In both cases, peptide and protein samples usually enter the sample preparation as aqueous solutions, which can contain varying amounts of an organic solvent, most often acetonitrile, methanol, or ethanol, and are usually acidified (pH 1–3). Acidification is, with only a few exceptions, a prerequisite for MALDI but not for ESI, which is far more flexible regarding the pH of the sample solution if appropriate (compatible) buffer systems and ion-pairing agents are used.

An important task of the MALDI sample preparation is to isolate and embed the analyte molecules in a solid environment of matrix molecules. According to the original method, the dried-droplet technique, which is still one of the most often used, this is achieved by mixing a small aliquot (0.5–1 µL) of a solution of the matrix compound in a mixture of acidified water and organic solvent with a similar volume of analyte solution on the sample support [63, 64]. During solvent evaporation, crystallization of the matrix commences and analyte molecules are trapped inside the growing crystals. This process is a key requirement for MALDI and highlights the important role of the sample preparation for its success. The matrix molecules are present in high molar excess compared to the analyte molecules, and absorb the laser light, enable desorption/ionization of the analyte molecules, and determine in the first place their internal excess energy, which determines their lifetime. If this is too short, depending on the m/z analyzer used, the analysis will fail or suffer from undesired early decay of the analyte ions. In ESI, in contrast to MALDI, the internal energy of analyte ions is determined by collisions with neutral molecules during transfer of the desorbed ions into the mass spectrometer.

Compared to ESI, MALDI is known to be more tolerant toward commonly occurring sample impurities such as salts and reducing or chaotropic reagents. Ionic detergents such as SDS, high molecular weight compounds such as NP40 or the Tween product family, and liquid nonvolatile additives such as glycerol, however, are well known to deteriorate the performance of MALDI. For this reason, depending on the sample preparation protocol used, these reagents can only be tolerated in moderate or very small amounts [36]. On the other hand, ESI benefits from the homogeneity of liquid samples and can be coupled online with reversed-phase (RP) LC, which eliminates most impurities that are known to interfere with ESI.

The main goal of the sample preparation protocol is to purify and enrich the analyte molecules and to deliver them in a format that best fits the ionization process. In addition, it should be simple, reproducible, and suitable for automation. For peptides, the best technique to purify them and recover them afterward in a small volume is RP chromatography. If not coupled with an LC system, small columns integrated at the outlet of a glass capillary or pipette tip have proven efficient for this task [65–68]. In this way, analyte loss due to surface adsorption is minimized and sample recovery is possible with only one microliter, which is sufficient for nano-ESI MS (see below) and more than needed for MALDI MS. The latter would benefit from even smaller volumes, the handling of which, however, can be difficult.

With respect to detection sensitivity and tolerance for sample impurities, the invention of nano-ESI, in which the sample is sprayed at nanoliters per minute compared to the microliters used before, has significantly improved the performance of ESI MS [69–71]. This is enabled by a much smaller orifice through which the charged sample liquid is sprayed, compared to conventional ESI (Fig. 3). As a consequence, the droplets formed are smaller and thus the series of events (solvent evaporation and droplet decay) is shorter until the analyte ions are airborne. The result is a significantly higher de-

Analysis of DNA-Binding Proteins by Mass Spectrometry

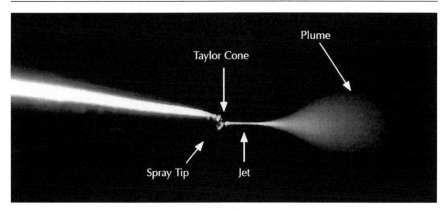

Fig. 3 ESI is a technique that uses high voltage to generate ions from an aerosol of charged liquid droplets. When utilizing flow rates from tens of microliters to milliliters per minute, aerosol formation must be assisted by pneumatic nebulization and/or by thermal heating to obtain a stable spray. This requirement is especially pronounced for highly aqueous liquids. When the flow rate is reduced to nanoliters per minute, referred to as nano-ESI, droplet formation occurs more readily, requiring only the applied voltage to generate a stable spray. The photograph shows the process, viewed through a high-powered microscope. As the liquid begins to exit the needle it charges up and assumes a conical shape, referred to as the Taylor cone, in honor of G.I. Taylor who described the phenomenon in 1964. The liquid assumes this shape because when charged up, a cylindrical shape can hold more charge than a sphere. At the tip of the cone, the liquid forms a fine jet, which then becomes unstable, breaking up into a mist of tiny droplets. Since these droplets are all highly charged with the same polarity they repel each other strongly. The introduction of nano-ESI has significantly improved the analysis of peptides and proteins by ESI MS. It is tolerant to a wide range of liquid compositions, and can even spray 100% water with a high degree of stability. The efficiency of ionization strongly improves as the flow rate is lowered because less volume of mobile phase passes through the emitter, producing smaller aerosol droplets. In addition, lower flow rates allow for longer analysis times, which is especially important when coupling nanoscale liquid chromatographic separation techniques online with ESI MS (Sect. 4.7). Taken from http://www.newobjective.com/electrospray/ with permission from New Objective, Inc., USA

sorption/ionization yield and a higher tolerance to nonvolatile sample contaminants [70, 71]. The difficulty in the manual handling of the thin and fragile glass needles required, as well as an overall lack of reproducibility and automation, has recently been overcome by the introduction of nano-ESI chips [72]. These contain nozzles optimized for nano-ESI and manufactured from silicon, which are now available in many different formats including arrays of 96 identical nozzles. As a consequence, fully automated analysis of hundreds of different samples by nano-ESI MS has become possible, which before was clearly a domain of MALDI-TOF MS. Another highlight has been the integration of RP LC separation in the outlet of a nano-ESI glass needle, eliminating any interface dead volume, and this approach has been further extended to the implementation of online 2D-LC (cation exchange followed by reversed-phase separation) [73, 74].

For MALDI, regarding detection sensitivity and automation, the invention of the thin-layer sample preparation technique brought about a first significant improvement for the analysis of peptides [75]. Instead of mixing small aliquots of matrix and sample solution on the sample support, first a small aliquot of the matrix CHCA dissolved in a fast evaporating solvent (acetone/water 99:1, v/v) is dropped on the support leading to the formation of a thin layer of microcrystalline CHCA. Then 0.5–2 µL of acidified sample solution is deposited onto this layer and allowed to dry. Before analysis, the dried sample is usually washed at least once with a larger aliquot (3–5 µL) of acidified water. Compared to the dried-droplet technique, where the analyte molecules are incorporated into the depth of the matrix crystals, with the thin-layer sample preparation method they are enriched in the upper molecular layers of the matrix crystals or adhere to their surface.

Another step forward was the introduction of prestructured sample supports (commercialized as AnchorChip™ technology). These have a strongly solvent-repellent surface equipped with a well-defined array of small hydrophilic spots (200–800 µm) acting as sample anchors. Each sample droplet contacts one of them and, during solvent evaporation, concentrates onto it enabling efficient concentration of the analyte molecules in the last step of the sample preparation. It has been shown that such supports can enhance the detection sensitivity of MALDI MS by one or even two orders of magnitude [76, 153, 222]. Besides the analyte molecules, however, all other, nonvolatile compounds are also enriched, which limits this approach to clean samples. For the analysis of peptides, this limitation was overcome by the CHCA affinity sample preparation technique, which combines the thin-layer sample preparation technique with prestructured sample supports and takes advantage of the recognition that microcrystalline CHCA has a high RP affinity and binding capacity for peptides [77, 78]. The CHCA affinity sample preparation technique integrates sample purification and concentration in the last step of the sample preparation and yields homogeneous samples at predefined locations and of predefined dimensions, which renders fully automated analyses straightforward (Fig. 4).

Surface-enhanced laser desorption/ionization (SELDI) is a proprietary technology, which significantly extends the above approach of surface-modified MALDI target plates [79–84]. It provides on-target enrichment and purification based on ion exchange, metal affinity, or normal-phase or RP chromatography as well as other, more specific interactions. SELDI is bound to a specific format and MS hardware and is mostly used for protein profiling in the context of clinical proteomics, where it has gained popularity. Some examples, however, also document its utility for affinity capture and subsequent characterization by MS of DNA-binding proteins [85, 86]. Until recently, with regard to resolving power and mass accuracy, the MS hard-

Analysis of DNA-Binding Proteins by Mass Spectrometry

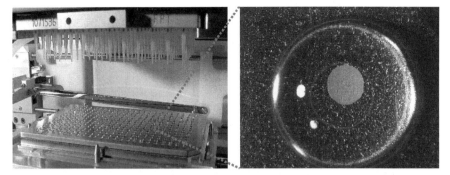

Fig. 4 CHCA affinity MALDI sample preparation of peptides. This technique takes advantage of prestructured sample supports (hydrophilic sample anchors surrounded by a hydrophobic support) and the observation that microcrystalline CHCA has a high RP affinity and binding capacity for peptides. It integrates sample purification and concentration in the last step of the sample preparation. Typically 0.5–2.0 µL of acidified sample solution (pH 1.5–2.5) is deposited onto one matrix spot measuring 400, 600, or 800 µm in diameter. Depending on the purity and concentration of the samples, they are either allowed to dry at ambient temperature (option 1) or removed after a defined incubation time, e.g., 3 min (option 2). In either case, all samples are washed once or multiple times with a larger volume of acidified water (3–8 µL) before they are analyzed. All these steps can be performed manually or automated using a pipetting robot as shown on the *left*. If the samples contain a lot of undesired contaminants that are difficult to completely wash away, option 2 is preferred. If their concentration is very low, the affinity purification yields benefit from longer incubation times because the samples' volumes continuously shrink over time until all solvent is evaporated. Therefore, if the contaminants can easily be washed away, option 1 is recommended because it provides maximum sample concentration and is easier to perform than option 2

ware that SELDI was bound to was not the optimal choice for peptide-based protein research. Fortunately, this has changed and the chances are fair that SELDI will soon also become a popular platform for protein identification following affinity purification. Unfortunately, their proprietary nature has, so far, excluded combining the advantages of the AnchorChip and the SELDI technology, which, no doubt, would be an attractive analytical union.

4
Mass Spectrometry Based Methods

For the identification and characterization of DNA-binding proteins, many different mass spectrometric methods are available, of which the most important are described and discussed in the following sections. Some of these methods have been specifically developed for this task, whereas others apply to proteins in general. Protein characterization, as it is understood here, not only refers to sequencing, modification analysis, and quantification, but also to exploration of protein higher-order structures, interaction networks, binding stoichiometries, surface topology, folding, and dynamics. For the identification and characterization of nucleic acids, other methods, optimized for this task, are available. Some of them are useful for the study of DNA-binding proteins. For instance, the primary structure of short oligonucleotides, including chemical modifications, can be determined by MS/MS and longer sequences can be determined by combining MS with limited exo- or endonuclease degradation. These applications of MS are not covered here but have been reviewed in detail elsewhere [87–89].

4.1
Protein Identification in Sequence Databases

For protein identification in large sequence databases, two fully independent and therefore complementary approaches have been developed: peptide mass fingerprinting (PMF, also called peptide mass mapping) and the correlation of peptide fragment ion data with known or predicted peptide sequences. PMF is based on the recognition that the molecular masses of a set of peptides of a proteolytic digest can uniquely identify that protein if a specific protease is used and its cleavage specificity is known and reliable [90–95]. The endoprotease used most often for this task is trypsin, which catalyzes exclusively cleavages at the C-terminal side of lysine and arginine residues, unless the next amino acid in the sequence is proline. In a database search, the measured masses are then matched against sets of expected masses calculated for all protein sequence entries. The search returns a list of candidates with the highest number of matching masses, and various algorithms are used to rank them and calculate the probability that the highest-ranking sequence entry is a true hit.

The concept of PMF has several distinct advantages compared to other protein identification techniques. For example, if the primary structure of a protein is subject to modifications (see below), with PMF only one or a few peptide masses will drop out of the identification while all the others will still match the correct protein sequence (Fig. 5). Furthermore, the detection sensitivity and mass accuracy are both significantly higher for peptides than for proteins, with the consequence that although the intact protein may not be detected in a sample it might still be identified by PMF. A limitation of the technique is that it assumes that the detected peptides are derived from a single protein. If more than one protein is present in the sample, the confidence of the search results decreases drastically. Therefore, the technique is most often used in combination with high-resolving protein separation techniques such as two-dimensional gel electrophoresis (2-DE) or affinity chromatography followed by gel electrophoresis or LC.

There are several strategies for protein identification using MS/MS data. Analogously to viewing a PMF as a fingerprint of the analyzed protein, a frag-

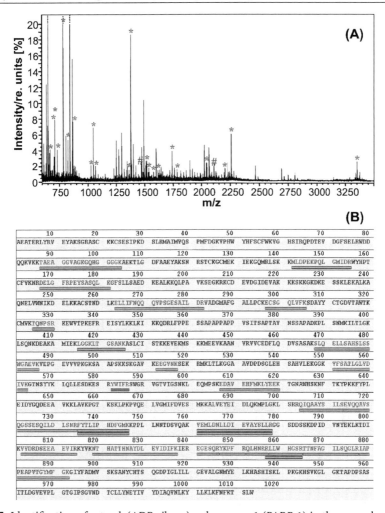

Fig. 5 Identification of rat poly(ADP-ribose) polymerase-1 (PARP-1) in the nonredundant UniProt/SwissProt protein database (sequence entry: P27008) by MALDI-TOF MS peptide mass fingerprinting. The protein was isolated from a nuclear extract by DNA-affinity purification followed by sodium dodecyl sulfate–polyacrylamide gel electrophoresis and then digested in situ with trypsin. **a** The recorded mass spectrum. *All peaks that were assigned to tryptic peptides of rat PARP-1 with a maximum relative error of 25 ppm. # In this case two peaks were assigned to the same peptide sequence containing one methionine residue, i.e., the unmodified and the singly oxidized molecular ions. Partial oxidation of methionine and tryptophane residues often occurs during the MALDI sample preparation and is taken into consideration when searching a protein sequence database and scoring the retrieved sequence entries. If both peaks are detected and the matched peptide sequence contains one of the two or both amino acid residues, this assignment is less likely to be a false positive hit than it would be if only one of the two peaks had been detected. **b** The sequence of rat PARP-1. Bars indicate the matched tryptic peptide sequence

ment ion spectrum can be regarded as a fingerprint of the fragmented peptide and thus be used directly for protein identification in sequence databases [96–103] (Fig. 6). The fragment ion masses determined for one peptide are compared to all sets of possible fragment ion masses calculated for each proteolytic peptide sequence, which matches the mass of the analyzed peptide

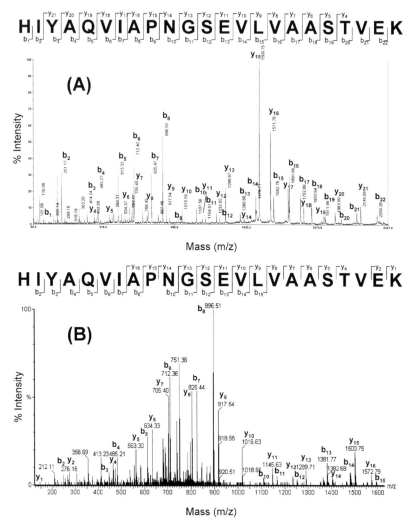

Fig. 6 Tandem mass spectra of a tryptic peptide from *Escherichia coli* ribosomal protein L18. **a** Spectrum produced from singly charged peptide molecular ions by a MALDI TOF/TOF mass spectrometer. **b** Spectrum produced from triply charged peptide molecular ions by an ESI Q-TOF mass spectrometer. Reproduced from [38] with permission from © American Chemical Society, 2006

molecular ion, for each protein sequence in the database. A second strategy is de novo sequencing with the goal of generating sufficient primary structure information from individual cleavage peptides to identify the corresponding protein sequence entry in the database. This is achieved by interpretation of the acquired fragment ion spectrum, e.g., by assigning m/z distances between signals to single amino acid residues [104].

Successful de novo sequencing requires spectra of high quality, ideally containing contiguous signal series representing the same ion type. These spectra, however, can contain signals of many types of fragment ions if different reactions contribute to their formation, rendering the interpretation difficult. Furthermore, the types and number of fragment ions formed strongly depend on the peptide's primary structure. Therefore, although the success rate of MS-based de novo sequencing of proteolytic peptides has increased impressively over the past few years, it is not yet routinely applicable as the only method for protein identification [71, 105–110]. Alternatively, when the sequence information generated only covers two to four amino acids, which is insufficient for searching large sequence databases, it can be combined with the molecular mass of the peptide, the cleavage specificity of the protease used, and the distance (in mass) of the determined sequence stretch from the peptide termini to form a so-called peptide sequence tag [7, 17, 111]. In contrast to short amino acid sequences alone, these tags provide highly specific input for database searches.

MS/MS experiments, especially when using CID, often yield only a few fragment ions of which one or two dominate the spectrum due to the presence of a preferred site of fragmentation, such as the peptide bond after an aspartic acid which is followed by a proline residue. Other examples are phosphopeptides, for which the first and highly favored fragmentation reaction is very often loss of the modification. In these cases MS^3 offers an attractive alternative to MS/MS to generate sufficient fragment ion data for protein identification [112]. When using ESI, instead of CID another possible option is to use ECD or ETD to induce fragmentation. In this case, the position of the captured or transferred electron (radical), and not the location of the weakest bonds of the molecular ion, determines in the first place the fragmentation pathway, with the consequence that both ECD and ETD are little sensitive to the presence or absence of secondary modifications [48, 50].

There are many other possible means to improve MS-based de novo sequencing of peptides. Obviously, if possible with the available MS instrumentation, combining different fragmentation techniques, e.g., CID and ETD, is one possibility to improve the sequence coverage (see above) and this can include one, two, or even more stages of ion isolation and fragmentation (MS^n) [48, 50, 112]. A different strategy is to label the peptide at one end or a specific residue with a chemical group (tag) that positively influences the fragmentation reaction or simplifies the data interpretation, or both [113–116]. The list of tags and labeling techniques that have been developed for this task is long and cannot be covered here. Some prominent examples are described and discussed in the context of protein quantification and labeling (see below).

A disadvantage of all the above approaches for protein identification, relying exclusively on fragment ion data, is that the identification usually covers only a small or tiny portion of the sequence of the identified protein. If no additional information is available, it is unclear whether the assigned full-length protein sequence or only a part of it was in the sample analyzed. One should keep in mind that with these experiments, each time only a proteolytic peptide is identified.

There are several strategies to enhance the confidence of identification results, and to specify in more detail which parts or possible variants of an assigned protein sequence were really in the sample. First of all, PMF and MS/MS-based protein identification are complementary methods that compensate each other's weaknesses. For instance, PMF data can be used to retrieve a list of possible protein sequence candidates, the matching proteolytic peptides of which are obvious candidates for subsequent verification or falsification by MS/MS. This approach is very powerful and straightforward, especially if the protein was isolated by 2-DE. In that case, the molecular mass of the protein as well as its isoelectric point can be estimated from the gel and be used as independent additional information to assist protein identification. Analogously, if the cleavage peptides are separated by RP LC before they are subjected to MS and MS/MS, their retention time provides additional valuable information for their identification [117–119].

If proteins have been isolated by affinity chromatography, it is useful, if possible, to analyze them directly by MS before proteolysis and to use these data to guide the next analytical means as well as the interpretation of the identification results. Another efficient means is to cleave the target proteins in separate experiments with more than one specific endoprotease, best two or more that have complementary cleavage specificity, e.g., trypsin, Lys-C, and V8 protease. In this way, several independent PMFs can be used for identification and more peptides are available for MS/MS analyses. Apart from these examples, any other information regarding the origin of the proteins to be identified, such as species of origin and their preferred cellular location, as well as their function, e.g., a known catalytic activity or binding affinity, can be helpful to control the quality of database search results.

4.2
Protein Sequencing

A dream of any protein researcher using MS is de novo sequencing of intact proteins in complex mixtures solely by MS^n. Although, using ESI-FT-ICR MS^n, nearly complete de novo sequencing has been demonstrated possible for a few model proteins, this approach has not yet found any broad application in protein research due to the enormous costs of the instrumentation, difficulty in the data interpretation, and the fact that its success depends much on the protein's sequence and structure, and therefore cannot be predicted [45, 120–125]. So far, the most successful approach for complete de novo sequencing of proteins has been to combine MS with complementary techniques, e.g., Edman sequencing. Obviously, the benefit of using MS in selecting peptide fractions for sequencing and to determine their position in the protein's primary structure is enormous and, no doubt, this approach has improved the quality, success, and turnaround times of de novo sequencing significantly [126–131].

Over the past 20 years, however, the strong advances in genomic and cDNA sequencing and protein sequence prediction have changed the situation dramatically. Today, instead of de novo sequencing a protein species, identifying it in a comprehensive sequence database is the obvious first attempt [7, 10, 17, 48, 130, 132–134]. If not successful, the next step is to search expressed sequence tag (EST) databases using all possible reading frames or even the entire genome of that organism [135, 136]. To be successful, this approach requires very specific search input data; best a set of high-quality MS/MS data or, even better, long sequence tags. If no or insufficient genomic data are available, at least partial sequence information can be generated with most state-of-the-art instrumentation by MS/MS-based de novo sequencing of a set of proteolytic peptides, and this information can be used to design oligonucleotide probes for screening cDNA or genomic libraries [71, 110].

Today, combining the activity of specific endoproteases with ESI MS/MS, the chances of determining 50% of a protein's sequence are fair if the quality of the recorded MS/MS data is good and if more than one enzyme is used. However, complete de novo sequencing is an art and anything close to 90% an impressive case. For this reason, the most promising approach to realize the ultimate goal of a comprehensive, error-free protein sequence database that covers many model species is to use MS to validate protein sequences that have been predicted based on genomic data, to eliminate sequencing errors, and to discover and describe possible modifications and sequence variants [137]. In fact, instead of de novo sequencing, the major task left for MS seems to be resequencing and modification analysis, which is described and discussed in the following section.

4.3
Protein Modification Analysis

Analysis of the primary structure of DNA-binding proteins with regard to modifications has become an ever-growing demand. Popular examples of PTMs are phosphorylation, acetylation, methylation, or ubiquitylation of individual amino acid residues. DNA-binding proteins, however, are also the subject of more exotic reactions, e.g., ADP-ribosylation catalyzed by the poly(ADP-ribose) polymerase (PARP) family. Prominent candidates are the histones, which are subject to all of these reactions and can harbor complex patterns of them, which led to the postulation of a histone code consisting of a series of site-specific modifications that encode information about chromatin structure and activity [138–141].

A second class of modifications is introduced on the transcript level, e.g., protein sequence variants that originate from mRNA splicing events. A third class occurs on the genome level, e.g., exchange of individual amino acids caused by single nucleotide polymorphisms. Finally, another category, which is not always easy to distinguish from the others, is modifications that can occur during the sample preparation [142]. Well-known examples are deamidation of asparagine and glutamine, carbamidoethylation of cysteine, and oxidation of methionine and tryptophane residues or proteolytic N- or C-terminal truncation of the protein's primary structure. For protein modification analysis, MS is the key technology for the still ongoing discovery of new modification reactions and target sites, as well as the readout of modification patterns and studying their interdependence.

The experimental strategies to detect and describe protein modifications by MS comprise up to four levels [143]. On the first, the molecular mass of the proteins is measured. If different from the value calculated for the unmodified sequence, this indicates a modification and the observed difference can be matched against a list of mass differences calculated for all known modifications to suggest a likely candidate. In some cases, if a specific modification is expected and the observed mass explains it, this information might suffice as confirmation. If none of the known modifications matches it, any further conclusion on this level is difficult or simply not possible as, most likely, several different modifications account for it. In many cases, however, the results of such analyses are more complex. For instance, instead of one signal, to which a molecular mass can be assigned, often a broad undefined signal, resulting from the overlay of many signals, or a series of fully or partially resolved signals or anything in between, is observed. In the worst case, the resolving power is insufficient to provide any useful information, which directly opens the gate to level two described below.

If signals are resolved, the molecular masses assigned to them can fall below or above the expected value and the latter can match one of them or none. Regular shifts can be very informative, e.g., a mass difference of 14, 16, or 42 Da matches the transfer of a methyl group, oxygen atom, or acetyl group, respectively. Triple methylation of a lysine residue, however, also matches an increase in mass by 42 Da and both possibilities are known PTMs of DNA-binding proteins. To distinguish which of the two is present, the mass difference needs to be determined with high accuracy (expected: 42.0470 Da for triple methylation and 42.0106 Da for acetylation), which today is possible with good ESI MS instrumentation if the protein is rather small (< 20 kDa). If not, that decision is open and needs to be addressed on the next level, where proteolytic digests are analyzed to determine which part of the protein's sequence is affected.

As the mass accuracy and resolving power is much higher for peptides than for proteins, the above question about the origin of an m/z shift of +42 can now be answered with many instruments including MALDI-TOF MS. On this level, even complex modification patterns that could not be separated before can be resolved and assigned to specific regions of the protein's primary structure (Fig. 7). As a next step, these peptides can be analyzed by MS/MS [50, 143, 144]. On this third level, individual residues are checked for a shift in mass that matches the observed deviation. If successful, these measurements assign modifications to individual amino acid residues, which is the natural endpoint of the analysis if their structure is known (Fig. 8). If not, sometimes the MS/MS spectra contain valuable information that can help to decipher the structure in question, e.g., fragment ion data that result from side-chain cleavages.

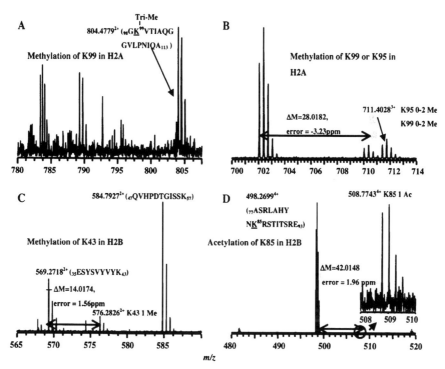

Fig. 7 Identification of novel posttranslational modifications of histones H2A and H2B isolated from calf thymus nuclei by FT-ICR MS analysis of proteolytic peptides generated by the digestion of histone H2A with pepsin (**a**) or trypsin (**b**), and of histone H2B with trypsin (**c**) and V8DE (**d**). Reproduced from [375]

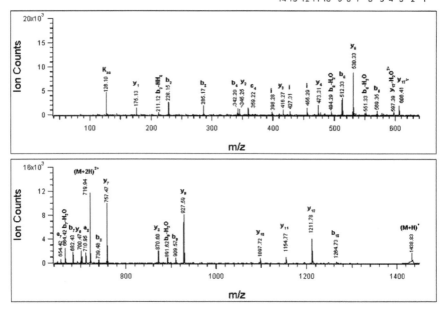

Fig. 8 Nano-ESI MS/MS spectrum of a doubly charged tryptic peptide (m/z 719.94) of histone H4 isolated from HeLa cells. Interpretation of the CID spectrum established the tetraacetyl-modified sequence covering residues 4–17 as GKacGGKacGLGKacGGAKacR, meaning that H4K5, 8, 12, and 16 are acetylated. Ions labeled as a, b, c, and their corresponding water or ammonia loss ions are N-terminal and those labeled as y and their water loss ions are C-terminal sequence ions. Ions labeled as i are internal fragment ions: m/z 356.20 corresponds to the fragment KacGGA; 427.31 corresponds to GGKacGL-28, KacGGKac, GKacGLG-28, GLGKacG-28, or LGKacGG-28; and 455.29 corresponds to KacGGKac, GKacGL, GKacGLG, LGKacGG, or GLGKacG. The signal at m/z 126.10 is the immonium ion of Kac. Reproduced from [235] with permission from © The American Society for Biochemistry and Molecular Biology, Inc., 2006

When MS/MS data are insufficient to clarify the structure of the modification, MS^3 or even MS^4 experiments represent the fourth level, where individual fragment ions are isolated inside the mass spectrometer, further fragmented, and the resulting ions analyzed to obtain more structural information [145]. This approach is especially promising if the individual amino acid residue that carries the modification or the modification itself can be isolated as a fragment ion for subsequent MS/MS analysis. An alternative is sometimes possible and often more successful, i.e., the use of specific enzymes or chemical reactions that release the attached chemical structure or parts thereof. These as well as the remaining peptide can then be analyzed by MS and MS^n to finish the analytical puzzle [143, 146–148].

Other important strategies focus on the enrichment of modified species or their efficient detection in complex samples based on the properties of a specific modification. This can be done before or after proteolysis with the consequence that the MS work starts on level one or two. Both approaches can be powerful and, because they are complementary, be combined to improve the analytical performance. Prominent examples of the enrichment of phosphorylated peptides are the use of immobilized antibodies, which specifically bind phosphorylated tryptophane or serine and threonine residues [144, 146, 149]. Because phosphorylation alters the net charge of the affected residues, its products can also be separated from unmodified molecules by ion-exchange chromatography. Another option is metal-affinity purification [150], which works best if applied after strong cation-exchange chromatography [112]. Acetylation of lysine residues is another good candidate for ion-exchange separation because it neutralizes their basicity.

Instead of separation, specific features of a modification can also be used for MS-based screening of complex samples. For this purpose, all ionized sample molecules are fragmented inside the mass spectrometer and the resulting daughter ions are scanned for the presence of a reporter ion (precursor ion scan) or an expected mass loss (neutral loss scan), which are specific for a certain modification [151]. Both techniques have proven powerful, especially for the rapid and sensitive detection of phosphopeptides [18, 144, 146, 152]. In this case, a neutral loss of 80 and 98 Da suggests the loss of HPO_3 and H_3PO_4, respectively, and the fragment ions m/z 79 (PO_3^-) and (or) m/z 97 ($H_2PO_4^-$) are specific reporter ions [18, 152, 153]. If that check is positive, the corresponding peptide is identified and further characterized by MS/MS or MS^3, in case fragmentation of the intact molecular ions is dominated by loss of the modification and the acquired MS/MS spectra therefore contain little sequence-specific information [112].

4.4
Analysis of Noncovalent Complexes

ESI can desorb and ionize noncovalent complexes and the ion desolvation and transfer conditions can be used to control their internal energy before m/z analysis [154–160]. This is necessary to enable the loss of all solvent molecules but not the interaction in view. Compared to ESI, MALDI has not been very successful for the desorption/ionization of intact noncovalent biological complexes and can therefore not be considered a serious alternative for this task. Meaningful ESI MS measurements on noncovalent complexes are not simple to conduct and managing the risk that the observed complexes, analyzed in vacuum in the absence of solvent molecules, might not reflect biological reality requires well-chosen control experiments and careful data interpretation [154, 156, 157]. As a rule of thumb, weak interactions that strongly depend on the physiochemical properties of the surrounding solvent molecules are no easy candidates for ESI MS, whereas strong interactions favor the analysis.

This means that for the analysis of DNA–protein complexes, the chances for success should be fair because their total binding energy is dominated by strong electrostatic interactions between the negatively charged phosphodiester backbone of the DNA and positively charged lysine and arginine residues of the protein component. This is supported by a number of examples that have been published and that document the feasibility and value of ESI MS for the study of noncovalent DNA–protein complexes [154, 160–163]. In addition to DNA molecules, such experiments can also include other proteins (multiprotein–DNA complexes) as well as potential cofactors such as cyclic adenosine monophosphate. Problems arise if the nucleic acid component is not well defined or long (see below). The obvious goal of the above measurements is, besides the detection of noncovalent interactions, the determination of binding stoichiometries in dependence on pH, buffer composition, salt concentration, and other factors. These can include different DNA probes (sequence dependence) as well as sequence variants of the proteins involved or the presence versus absence of a specific modification (e.g., phosphorylation). Furthermore, information about the structure, stability, and dynamics of a complex can also be derived from changes of the observed charge state distributions and binding stoichiometries in dependence on the temperature of the sample solution [154, 160, 161]. Such experiments are especially informative if combined with hydrogen/deuterium exchange experiments (see below).

It is noteworthy that of the more than 500 publications that report on ESI MS of noncovalent protein complexes, only about 2% cover protein–nucleic acid interactions [154]. This discrepancy reflects well the difficulties encountered when analyzing protein–nucleic acid complexes by ESI MS. These include the high salt concentrations in the sample solution necessary to stabilize specific interactions, as well as nonspecific cation binding to the negatively charged backbone of the DNA molecules involved. As a result, peak broadening is a common observation and the accuracy of mass assignments can be poor.

Another problem that needs to be dealt with is the strong local analyte enrichment that results from the loss of solvent molecules by evaporation during the process of ESI. As a consequence, multiply positively charged proteins bound or not bound to highly negatively charged nucleic acid oligomers are forced close to each other, which increases the risk of nonspecific interactions (false positive results). This explains why the reality looks somewhat different than the above rule of thumb suggests [154]. Nevertheless, the published examples document the enormous potential of the technique. These include the detection and characterization of individual biomolecular interactions, as well as the analysis of complex macromolecular machines such as ribosomes by ESI MS [155] (Fig. 9).

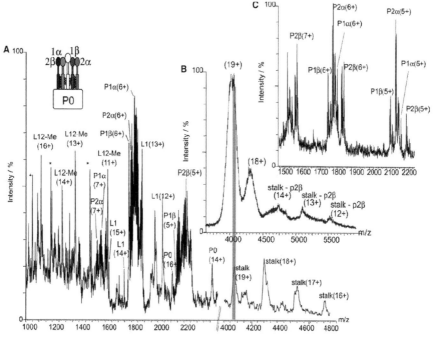

Fig. 9 Mass spectrum of intact ribosomes from *Saccharomyces cerevisiae*. **a** A series of peaks were observed, which were assigned to the stalk proteins P1a/b, P2/b, and P0, large ribosomal proteins L1 and L12, and a stalk pentamer (P1/P2)$_4$P0. The peaks corresponding to the stalk pentamer are magnified by a factor of 2.5. **b** Tandem mass spectrum of the 19+ charge state assigned to the pentamer reveals charge stripping as well as complexes from which a single protein has been removed. The region of the m/z scale shaded *gray* represents the isolation of the single charge state for tandem mass spectrometry. **c** Low m/z region of the tandem mass spectrum showing the preferential release of the P2 proteins from the stalk complex, consistent with their location on the periphery of the pentamer. The schematic representation of a model of the stalk pentamer showing the relative locations of the P1/P2 proteins and their interactions with P0. * denotes other members of the L12 charge state series that have not been labeled due to lack of space. Reproduced from [155] with permission from © Federation of European Biochemical Societies, 2006

4.5
Protein Footprinting, Cross-Linking, and Surface Labeling

Combined with specific enzymatic or chemical reactions, both ESI and MALDI MS can be used to generate structural information about protein interaction complexes. The common principle across these experiments is that MS is used to monitor or read out the final result of a reaction that degrades or chemically modifies the interaction complex under nondenaturing conditions. Typical control experiments include the same reaction performed (a) in the absence of DNA, (b) using different DNA sequences, and (c) applying denaturing instead of nondenaturing conditions.

Comparing the mass spectra of the interaction experiments with those acquired from control experiments reveals differences that can be related to the structure of the interaction complex. Depending on the reaction, intact proteins, proteolytic peptides or both are simultaneously analyzed by MS and, optionally, MS/MS. If the sample is too complex for direct analysis, a broad range of additional means of separation are available, e.g., electrophoresis, LC, or affinity capture, which can all be efficiently combined with ESI or MALDI MS (Sects. 4.7 and 4.8). A major challenge for all three techniques described below is the data interpretation. This concerns less the identity of the resulting products than the molecular puzzle they create. Relating the observed differences between sample and control experiments to the structure of the interaction complex can be difficult, and great care is recommended before the data in hand are considered evidence for the existence of a certain structural element.

4.5.1
Protein Footprinting

The principle of protein footprinting relies on differences in the enzymatic proteolysis rate of proteins that closely interact with other molecules or not. The conceptual idea is that certain parts of their primary structure are protected (shielded) from proteolysis by close interactions with other molecules, with the consequence that the protease used will not or less efficiently catalyze cleavages within these regions (Fig. 10). As any cleavage, independent of where, can destabilize the interaction complex because it shortens or disrupts the primary structure of the protein, a major risk for protein footprinting is that the structure in question is quickly altered during the course of the reaction. Whether this happens early on, at a late state or not, is usually unknown. Therefore, the reaction is monitored over time, or the concentration of the protease is varied in parallel experiments [164–168].

For protein footprinting, an important advantage of studying DNA–protein versus protein–protein interactions is that the DNA is unaffected by the activity of the protease, and that its length, composition, and sequence can be changed in any way without the risk of unintentionally altering the proteolysis pattern. Furthermore, DNA molecules generated by PCR or chemical synthesis can easily be equipped with a broad variety of functional groups, optimized for specific tasks such as immobilization, affinity purification, or detection. This enormous flexibility, generally not available for proteins, enables informative controls and renders systematic follow-up experiments straightforward, e.g., changing nucleotide identities in order to study their effect on the structure and stability of the complex.

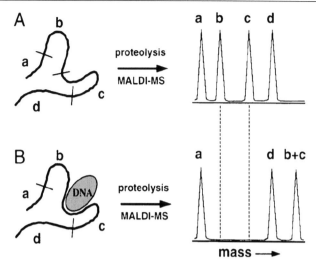

Fig. 10 Illustration of MALDI MS-based protein footprinting. The figure compares peptide mass fingerprints obtained by enzymatic proteolysis of a hypothetical DNA-binding protein, in the absence and presence of its target DNA. A *thick black line* traces the polypeptide backbone of the protein containing three proteolytic cleavage sites. **a** In the absence of DNA, the protein is proteolytically cleaved into four peptide fragments (*a*, *b*, *c*, and *d*). The mass spectrum of the resulting digest shows four peaks with masses that can be determined with sufficient accuracy to unambiguously identify the four peptide fragments. **b** In the presence of DNA, one of the three potential sites of proteolysis becomes protected following DNA binding. Under these conditions, the protein is cleaved into three fragments (*a*, *d*, and *b* + *c*) that can be identified in the mass spectrum. Because the site between *b* and *c* is protected, a single peak, corresponding to the combined fragment *b* + *c*, appears at higher mass. Reproduced from [165] with permission from © The Protein Society, 2006

Protein footprinting is not a new technology and various attempts using gel electrophoresis or LC, combined with Edman sequencing, date back long before ESI and MALDI. Its applicability, however, no doubt benefited enormously from the introduction of ESI and MALDI MS [164, 169]. This concerns the amounts of sample material required, the workload, and the analysis times. For instance, using MALDI-TOF MS, protein footprinting experiments can be accomplished within a day, rather than weeks, and instead of nanomoles only a few picomoles of the interacting molecules are required [164–168, 170–172].

The interpretation of protein footprinting data can be simplified if the DNA probe is immobilized on small magnetic particles, e.g., by biotin-streptavidin interaction. In this case, all reaction components bound to the beads and those that are not can be separated at any time simply by placing the reaction tube close to a strong magnet. To monitor the footprinting reaction over time, at defined time points a small aliquot (1–2 µL) of the reaction mixture (suspended beads) is removed and transferred to a separate vial. The contained beads are collected at the wall as described and the supernatant is transferred to another vial and then acidified to stop the enzymatic reaction. The collected beads are washed with digestion buffer and the bound protein components are then extracted with an acidic buffer or MALDI matrix solution. Both solutions are later analyzed by MS. The advantage of this strategy is that the protein degradation products that still bind to the DNA are separated from those that have lost this ability, before they are identified by MS. This information simplifies the data interpretation considerably [173].

Limitations of protein footprinting arise from the fact that many DNA-binding proteins only interact with the target DNA in concert, e.g., as homo- or heterodimers, or require other cofactors to form a stable complex. In all these cases, the challenge is to differentiate protein–protein or protein–cofactor from protein–DNA interactions [170–172, 174]. Although potentially a powerful tool, especially if combined with advanced software, so far, due to the above problems, protein footprinting has not become a popular, widespread technique. In contrast to this, the complementary approach of DNA footprinting is used routinely by many scientists to locate *cis* regulatory DNA sequence elements to which transcription factors bind specifically. MS, however, has not seriously been combined with this technique, simply for the reason that the performance of the established, electrophoresis-based techniques for analyzing DNA sequence ladders is not matched by any MS-based method currently available.

4.5.2
Cross-Linking

Cross-linking of peptide–DNA and protein–DNA complexes is an old, well-established technique that dates back to the early 1960s [175, 176]. Main applications have been the determination of binding stoichiometries, e.g., by subsequent gel electrophoresis, and identification of close neighbor residues that were linked by Edman sequencing. The combination with MS took another 30 years to take off [177, 178], which dates back to shortly after the new ionization techniques of ESI and MALDI were introduced. Since then, the performance and applicability of this analytical alliance has grown continuously [179] and today, no doubt, provides a powerful tool for the characterization of DNA-binding proteins.

Covalently linking protein–DNA complexes can be accomplished by a broad range of chemical reagents including small and simple structures, e.g., formaldehyde [180, 181]. Most experimental strategies, however, focus on light-induced cross-linking of amino acid residues to close neighbor nucleobases, which are good candidates for this approach. An obvious advantage of this strategy is that it does not require additional molecules to link residues in close vicinity. The short distance over which covalent links are formed this way, however, can also be considered a limitation. Controlling this parameter is the domain of specific, bifunctional chemical reagents that can be designed to bridge defined, larger distances [169, 182, 183]. This approach, however, is limited to residues that the reagent used can access from the outside, which combines cross-linking experiments, discussed here, with the strategy of surface labeling described below.

A severe problem of photochemically cross-linking native protein–DNA complexes, requiring light in the range of 250–280 nm, is the poor reaction yields typically observed for the available photosensitive groups as well as undesired photodamage, e.g., photocleavage or oxidation. Therefore, to enhance the yield and enable site-directed cross-links, specific nucleobases are often exchanged by synthetic derivatives, which are far more reactive than their natural analogs and absorb light at wavelengths above 300 nm, thus avoiding photodamage. Examples are tether-bound psoralen [184], benzophenone derivatives [185], and azido-, thio-, bromo-, and iodo-substituted nucleobases, e.g., 4-thio-, 5-bromo-, or 5-iodouridine [186, 187]. The latter group of analogs is often preferred because they are readily available and they minimize structural distortions of the native interaction. An important advantage of using nucleotide analogs is that individual nucleobases can be separately tested for their cross-linking yields, which renders obsolete the need for identifying which nucleotide was linked to a specific amino acid residue. As for protein footprinting, the flexibility in synthesizing the DNA oligomers available today significantly simplifies the study of protein–DNA interactions by cross-linking experiments.

The MS analysis of protein–DNA cross-linking products can comprise three levels. Firstly, MS can be used to determine cross-linking stoichiometries. This requires that the DNA is not too long or is shortened to a defined length before the products are analyzed by MS [179, 188]. The second level aims at assigning binding regions (domains). For this purpose, aliquots of the reaction products are incubated with a specific protease and the resulting cleavage products are analyzed by MS as described. Optionally, the DNA molecules can also be digested with the aid of an enzyme before or after that analysis (Fig. 11). This is usually a nuclease with the aim of reducing the size of the cross-linked DNA to small oligomers or individual nucleotides, which can be further degraded to nucleoside residues with the aid of alkaline phosphatase. Finally, at the third level modification analysis by MS/MS is used to identify exactly which amino acids were cross-linked to which nucleobases.

Compared to protein footprinting, the interpretation of cross-linking data is usually more straightforward, especially if the question of which nucleotide was linked to a certain amino acid residue is taken care of by the above nucleobase derivatives. The problems that can arise and limit the success of MS are directly connected with those well known for the analysis of nucleic acids (see above). They strongly increase when long instead of short oligonucleotides

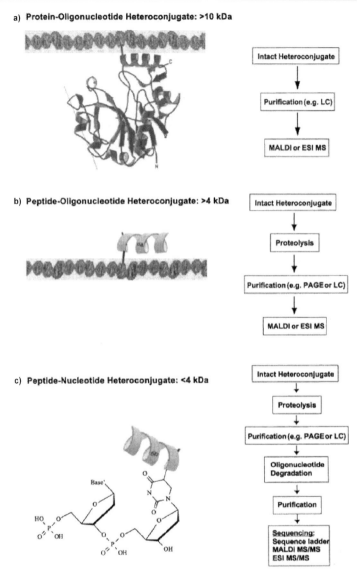

Fig. 11 Various types of cross-linked polypeptide–oligonucleotide complexes (*left panel*) and the analytical strategies to analyze them (*right panel*). Heteroconjugates are covalently coupled hybrids generated by cross-linking that can be analyzed at various stages. **a** The intact protein–oligonucleotide is analyzed with or without prior purification. At this stage, in the sample preparation and subsequent analysis the heteroconjugate is usually treated as a protein-like analyte. **b** After proteolytic degradation of the protein moiety, a peptide–oligonucleotide hybrid is obtained, which is analyzed as an oligonucleotide-like analyte. **c** After degradation of the protein as well as the oligonucleotide moieties, the analyte resembles a modified peptide, which is isolated and subsequently analyzed. Reproduced from [179] with permission from © Wiley Periodicals, Inc., 2006

are part of the sample. These require special treatment during the sample preparation and mass spectrometric analysis [78, 87, 88, 189, 190].

4.5.3
Surface Labeling

The third technique, surface labeling, uses a chemical reaction to covalently modify specific amino acids, e.g., lysine, tryptophane, or cysteine residues, which are exposed at the surface of the interaction complex. The reaction conditions are optimized such that only these residues are modified and not those hidden in the interior of the complex. Besides nondenaturing reaction conditions a second, equally important requirement is that the reagent used is too large to penetrate the surface of the interaction complex, but not too large to enable access to residues exposed within grooves or open pockets. In fact, the size (cross section) of the reagent is an important analytical parameter that determines the resolution of the surface accessibility experiment [191–194].

The task of MS is modification analysis as described above, with the goal of identifying the labeled residues. This work can be assisted by the design of the labeling reagent. For instance, if it includes a biotin tag, affinity purification using streptavidin or avidin columns or coated magnetic beads can be used to enrich, after enzymatic proteolysis, the labeled peptides, which greatly simplifies their analysis by MS. Problems in the performance of surface labeling experiments arise from solubility issues (reagent versus interactants) as well as maintaining the integrity of the interaction complex during the course of the reaction [191, 193, 194].

4.6
Hydrogen/Deuterium Exchange

Hydrogen/deuterium exchange (HX) coupled with MS (HX MS) is a powerful analytical tool to study the structure, dynamics, and assembly of all kinds of proteins as well as their interactions with other molecules. The method is based on the recognition that labile hydrogen atoms of a protein will exchange with deuterium atoms in the presence of D_2O [195–201]. HX is a well-established technique that can be combined with NMR [202], MS [199–201], and resonance Raman spectroscopy [203]. The most important advantage of using HX MS versus HX NMR is the high detection sensitivity and short analysis times of MS, which makes it possible to conduct a series of experiments with a few picomoles of sample in a single day. Other advantages are that MS is not limited to protein size as is NMR (approx. 40 kDA). With respect to their performance and requirements, however, these techniques should be considered complementary rather than competitive, e.g., a structure established by NMR (or X-ray crystallography) is a good starting point (reference) to study protein dynamics, assembly, and interactions by HX MS.

Proteins contain different kinds of exchangeable protons, which can be classified as fast and slow exchanging [199, 200]. All those bound to oxygen, sulfur, or nitrogen atoms are mobile and exchange too fast to be measurable by any isotope exchange method, except those bound to backbone amide groups, which exchange far more slowly and are the target of HX experiments. This reaction is performed under physiological conditions and is usually base-catalyzed (OD$^-$). Those hydrogen atoms directly bound to carbon are usually not exchanged without a special catalyst (harsh conditions). The resulting increase in mass, and thereby the number of exchanged hydrogen atoms, can be determined by MS and correlated with the expected maximum number possible, derived from the protein's primary structure.

Systematic studies have shown that individual exchange rates of amide hydrogen are in the first place dependent on solvent exposure and whether the hydrogen atom is involved in hydrogen bonding or not [199, 200]. Under physiological conditions, any backbone amide hydrogen in short peptides is usually replaced by a deuterium atom within 1–10 s, whereas in folded proteins the exchange of shielded amide hydrogen atoms can require many days or months. These variations reflect the diversity of their local environments with regard to solvent exposure and hydrogen bonding, and can be analyzed by MS and related to a known structure model or used to generate de novo structure information as outlined below.

The exchange of amide hydrogen atoms buried in the core of a protein or involved in hydrogen bonding requires structural changes that can be classified as local, short-living events such as thermally induced protein motions ("breathing"), which involve small atomic movements, and longer-living events that are associated with rearrangements of larger segments or global unfolding which both involve cooperative movements of many atoms over longer distances (several Å) [200]. In the first case, diffusion of OD$^-$ and D_2O to the exchange site is thought to determine the reaction rate. These exchanges can be very local with the consequence that individual amide hydrogen atoms can be exchanged while their next neighbors are still protected. The kinetics of these exchange reactions can be envisioned as involving many rapid and random short visits to a state capable of exchange, while the probability of exchange during each visit is small (EX2 kinetics).

In the second case, sudden exposure to D_2O over a longer time and disruption of hydrogen bonds enable exchange at groups (series) of amide bonds rather than individual sites, and the kinetics of these reactions can be compared with a long visit to a state capable of exchange (EX1 kinetics). Under physiological conditions native (folded) proteins are all the time subject to EX2 whereas EX1 exchanges are usually rare events, which, however, can be experimentally induced, e.g., by applying denaturing conditions [199]. The abundance of EX1 events over time can be measured and used to classify a structure as rather static (few changes) or dynamic (flexible). Differentiation of EX1 versus EX2 is based on the assumption that EX1 affects series of amide hydrogen bonds whereas EX2 affects individual sites. In reality, of course, this is not always possible as "thermal breathing" and folding events are interconnected.

HX MS is an efficient technique to study protein structure and dynamics, their assembly, as well as higher-order noncovalent complexes. It has, for instance, provided detailed insights into the folding of denatured polypeptide chains to functional proteins, the function of chaperones in assisting this process, and the structural changes enzymes undergo upon binding of a substrate or cofactor molecule [199, 200, 204–206]. The obvious analytical challenge is accurate measurements of HX, best at the level of individual amino acid residues, without affecting the current status of exchange. For HX MS this means that the sample preparation and its analysis are not allowed to induce significant additional in-exchange of deuterons or back-exchange of deuterons by hydrogen atoms. This is possible, because HX can be slowed down drastically by reducing the temperature to near 0 °C and the pH (pD) to 2–3, where the reactivity (catalysis) of this reaction reaches a minimum [199, 200, 207, 208].

For both techniques, ESI and MALDI MS, sample preparation protocols have been developed that meet these requirements. This also includes optional enzymatic cleavages using the endoprotease pepsin, which has high activity at the required low pH, as well as LC separation of the resulting peptide mixtures. In this way, it is possible to monitor HX over time as an increase in molecular mass of the intact protein (total HX) as well as of its defined segments (peptides). Ideally, this would be routinely further extended to individual amide bonds by the use of MS/MS. This, however, seems to be impossible by CID because, along with the increase of the internal energy, amide-bonded H and D are scrambled along the backbone of the peptide molecular ion and the information looked for is lost before bond breakage occurs [209, 210]. Current studies aim at overcoming this limitation by the use of ECD or ETD, which are known to induce instant decay upon electron capture [53, 56–58, 211]. If successful, this would significantly extend the performance of HX MS.

Other limitations of HX MS arise from the low specificity of the endoprotease pepsin, which preferably catalyzes cleavages at hydrophobic residues, and the lack of good alternatives [199, 200]. The main problem is not identification of the cleavage peptides, which is achieved by accurate mass determination and MS/MS, but the observation that for many proteins peptic peptides can be fairly long (e.g., 20 amino acid residues), which means poor HX MS resolution. Nevertheless, the benefits of combining HX with MS are convincing, especially its unsurpassed sensitivity, ease of use, and speed, with the consequence that its popularity has grown continuously over the past few years, and that fully automated workstations have been constructed that cover all aspects of the necessary sample preparation [200, 212].

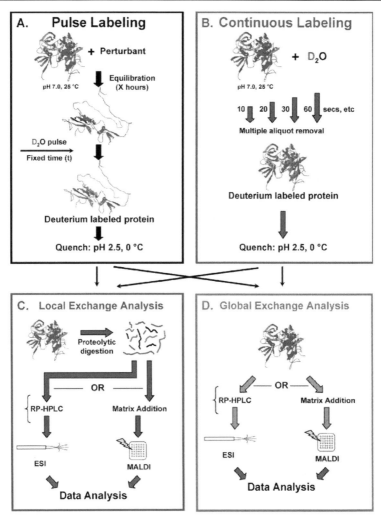

Fig. 12 Overall scheme for HX MS experiments. **a** Pulse labeling. After a protein has been exposed to a perturbant, e.g., chemical denaturant, heat, pH, etc., unfolded regions (*gray*) become labeled with deuterium (*red*) during a quick pulse of D_2O (typically 10 s). Deuterium exchange is quenched by reducing the pH and temperature. **b** Continuous labeling. D_2O buffer is added to a protein (in H_2O buffer) such that the final D concentration is > 95%. At defined time points, an aliquot of the labeled protein is removed from the original tube and mixed with quench buffer to reduce the pH and temperature. **c** Localized exchange information. Quenched samples (from part a, part b, or both) are digested with pepsin and the resulting peptides are analyzed by LC-ESI MS or MALDI MS. The resulting data analysis provides information on deuterium exchange in short fragments of the peptide backbone. **d** Global exchange information. Quenched samples (from part a, part b, or both) are directly analyzed by MS. The data provide a global picture of how the protein behaves in D_2O. It is often recommended that part d be performed prior to part c. Adapted from [199] with permission from © Wiley Periodicals, Inc., 2006

To measure the incorporation of deuterium, the sample must contain high doses of D_2O, which usually requires dilution or buffer exchange. Dilution is the easiest and fastest method but only attractive if the sample molecules are available in high concentrations. Thanks to the high detection sensitivity of ESI and MALDI MS, this is not a rare exception. However, in many cases dilution is not option and, instead, a rapid buffer exchange is performed, e.g., using a small gel filtration spin column or a size-exclusion membrane.

There are two alternatives to perform HX experiments, i.e., continuous and pulse labeling [199], of which the former is easier and the most often used (Fig. 12). In continuous labeling experiments HX is unidirectional (no back-exchange), with the consequence that the deuterium content of the participating proteins at any point in time during the course of the reaction integrates HX events. This information, quickly accessible by MS, is well suited to monitor slow unfolding transitions, the majority of unfolding events in proteins, in response to experimental conditions.

Compared to continuous labeling, pulse labeling experiments are more difficult to conduct (rapid exchange of D_2O and H_2O) and therefore less common. The advantage of this approach is that instead of integrating HX events, snapshots of exchangeable amide hydrogen are recorded. Due to back-exchange between two pulses, deuterium incorporation only accumulates if protected from back-exchange. Pulse labeling HX MS has been used to identify protein folding mechanisms and to assign kinetic intermediate states, which are difficult or impossible to detect by continuous labeling experiments.

4.7
Combining MS with Liquid Chromatography and Electrophoresis

LC and electrophoresis are the two most important techniques for separating the components of complex peptide and protein samples, respectively. While for proteins no LC separation technique has the resolving power of 2-DE [213–218], for peptides 2D-LC is the method of choice for separating very complex samples, which most often comprises fractionation on a strong cation-exchange column in the first dimension followed by RP separation in the second [9, 73, 74, 219–221]. The latter has high resolving power for peptides, provides strong enrichment, affords efficient sample desalting, and can be operated with aqueous–organic solvents directly compatible with ESI and MALDI. To optimize analyte recovery in the low femtomole range, modern LC systems, dedicated for this task, minimize all contact surfaces and dead-volumes and operate at flow rates in the range of nanoliters per minute (nano-LC). The eluent is most frequently analyzed online by ESI MS, but offline combinations with MALDI MS are catching up in popularity for good reasons (see below).

2-DE combines two efficient protein separation techniques, i.e., separation based on charge state in the first dimension followed by separation based on size in the second. As a consequence, for proteins the resolving power of 2-DE is much higher than that of any one-dimensional separation technique available. For peptide separation by 2D-LC the situation is different. In this case only the second-dimension separation is very powerful. The first-dimension fractionation by charge suffers from the fact that most proteolytic peptides, in contrast to proteins, accommodate at any pH maximum three or four charges. This means that the gain in resolving power is only moderate when the sample is fractionated by cation exchange before separation by RP LC. The reason why this is done anyway is that the two techniques can easily be combined online enabling efficient, fully automated workflows. It is well known that the combination of other LC separation techniques, e.g., hydrophilic interaction with RP or two times RP at different pH values using different ion-pair reagents, are superior in resolving power to cation exchange followed by RP LC. A major disadvantage of these combinations with regard to sample loss and automation, however, is that their elution conditions are in no orientation compatible with their sample loading conditions, which renders necessary drying and dissolving again each fraction collected in the first dimension along with an inevitable loss of sample.

With respect to resolving power (theoretical number of plates) and separation time, no doubt, capillary electrophoresis (CE) is the ultimate separation technique for complex peptide samples, and its combination with ESI MS online as well as MALDI MS off-line has been demonstrated many times [222–229]. The main reason why CE-MS, in contrast to nano-LC-MS, has not become a widespread method for protein and peptide analysis is the maximum total sample volume that can be separated by CE. In contrast to nano-LC, where many hundred microliters of dilute sample can be loaded without compromising separation power, the performance of CE directly depends on the sample volume and works best if only 50 nL or less is loaded. Recently, however, it has been realized that this requirement of CE is perfectly matched by nano-LC, which provides efficient sample concentration, and that the two techniques can be combined online upfront ESI or MALDI MS. For this purpose, a microfluidic chip was developed that enables, on demand, on-line transfer (loading) of nano-LC fractions to an orthogonal CE separation channel, the effluent of which is either analyzed online by ESI MS or off-line by MALDI MS [230–232].

With this setup, nano-LC provides a first dimension of separation during which the sample molecules are concentrated from a large volume, e.g., a few hundred microliters, to only a few hundred nanoliters or less. Each LC peak or a fraction of it is then separated in a short time (a few seconds) by CE and afterward analyzed by MS and MS/MS. The above 2D combination of nano-LC and CE in a chip format ready to combine with ESI or MALDI MS is a new development that has raised considerable interest. For the time being, the technique certainly looks promising but, unfortunately, too little data and reports are yet available that allow here to judge whether it has the potential for future widespread application or not.

Although 2-DE provides the highest separation power available for proteins, its applicability to the fractionation of DNA-binding proteins is limited. In the first place, this is due to the very low abundance of many DNA-binding proteins, which excludes their detection by 2-DE without previous efficient enrichment. This is not easy and usually requires target tailored affinity purification, which in return, if successful, renders the resolving power and effort of 2-DE unnecessary (Sect. 4.8). Other limitations arise from the observation that both very basic and very small proteins are not well appreciated by 2-DE. Instead, these proteins are optimal candidates for isolation by LC using different separation modes in series, e.g., size-exclusion, ion-exchange, and RP chromatography. The latter two are very popular for the separation of histones including some of their many possible PTMs, especially acetylation of lysine residues (Sect. 4.3) [233–238].

As the analysis of proteins by MS is more informative on the level of proteolytic peptides than proteins, and these are easy to recover from a gel matrix, gel electrophoresis and LC are frequently combined by in-gel proteolysis of the separated proteins (see below) (Fig. 13). In contrast to 2-DE, 1-DE, especially sodium dodecyl sulfate–polyacrylamide gel electrophoresis (SDS-PAGE), is routinely used for the separation of DNA-binding proteins. This is documented by the fact that for the identification of low abundant proteins, which bind to specific DNA sequence elements, among many other analytical strategies explored, affinity purification followed by SDS-PAGE followed by MS has proven to be the most successful [239–245]. SDS-PAGE is efficient in separating proteins by size, can concentrate individual fractions in a few microliters of gel matrix, and, regarding their physiochemical properties, accepts a broad spectrum of protein species. For the analysis of low abundant proteins, another important advantage of this separation technique is the SDS loading buffer, which is very efficient in solubilizing (recovering) tiny amounts of protein material. Finally, compared to 2-DE, SDS-PAGE is a simple, low-cost technique available in any molecular biology laboratory.

As complex mixtures of DNA-binding proteins can also be directly analyzed (displayed) by MS, best with MALDI-TOF MS [239], it is interesting to compare this approach with SDS-PAGE. The clear advantages of MS are accurate protein molecular mass determinations, higher resolving power, and much higher detection sensitivity for small proteins (< 10 kDa), which are usually underrepresented by SDS-PAGE. Disadvantages are that the resolving power and detection sensitivity both quickly decline with increasing molecular mass, and that none of the available protein solubilizing buffers compatible with MS is nearly as good for protein recovery as SDS-PAGE loading buffers. Furthermore, complex mixtures are often not well represented

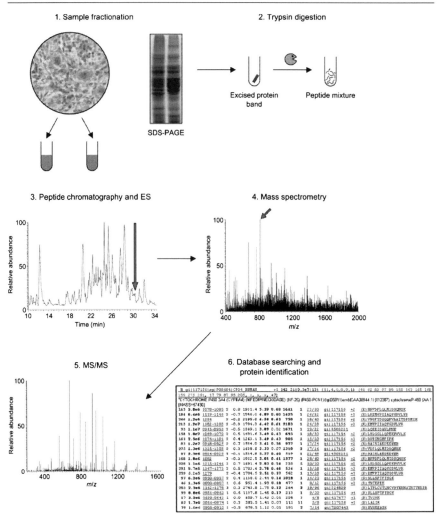

Fig. 13 A generic MS-based proteomics experiment, consisting of six stages. Proteins for analysis are first isolated from cell or tissue lysates by fractionation methods, often including SDS-PAGE as the final protein separation step. Proteins are then enzymatically proteolyzed in situ and separated using one or more stages of LC. The LC eluate is directed into an ESI mass spectrometer, and the peptides are analyzed first by MS and then by MS/MS. The recorded peptide MS/MS spectra are searched against a protein database using a search algorithm, which assigns peptide identifications based on match criteria. Reproduced from [10] with permission from © Birkhäuser Verlag, 2006

by MS (unpredictable discrimination of protein species). This observation is not common for SDS-PAGE. In addition, instead of declining, the detection sensitivity of SDS-PAGE, with respect to the number of molecules, steadily increases with increasing protein size. The inevitable consequence of the above comparison is that for differential display of protein mixtures, MS and SDS-PAGE complement each other well.

Combining 1- or 2-DE with MALDI or ESI MS is typically achieved by excising stained fractions out of the gel, washing them, in situ proteolysis of the isolated proteins, extraction of the released peptide mixtures, and optionally purifying and concentrating them by RP chromatography. Today, for direct analysis, with respect to detection sensitivity, throughput, and automation, MALDI-TOF MS is the preferred technique and ESI MS is used when additional MS/MS data are required. If the peptides are separated by RP LC before they are analyzed by MS, the situation is different and ESI is usually the first choice and MALDI is the technique that enables acquisition of additional MS/MS data (see below).

Recent developments have made off-line coupling of MALDI MS with LC an attractive alternative to online LC-ESI MS [246–254]. The off-line nature of LC-MALDI offers some distinct advantages compared to online LC-ESI for MS/MS analyses. With the latter ionization technique, the time available for MS/MS is limited to the time width of the corresponding chromatographic peak. For the analysis of complex peptide mixtures, where many compounds coelute, this time interval is often too short to analyze all of them by MS/MS. This time constraint does not apply to MALDI. After initial MS analysis, which only consumes a tiny part of the sample, the remainder can be analyzed without any time constraints, enabling efficient, data-dependent MS/MS analyses.

To evaluate the performance of MALDI versus ESI for the LC-MS/MS analysis of complex mixtures of proteolytic peptides of DNA-binding proteins, in one study all *E. coli* proteins that bind to single-stranded (ss) DNA were affinity purified, digested with trypsin, and aliquots of this mixture were separated by RP LC. The effluent was analyzed three times online with an ESI quadrupole-TOF hybrid mass spectrometer and three times off-line with a MALDI tandem-TOF (TOF/TOF) mass spectrometer. An important result of these experiments was that the two LC-MS techniques are highly complementary [38]. It was observed that ESI tends to favor the identification of hydrophobic peptides, whereas MALDI tends to lead to the identification of basic peptides and those that contain aromatic amino acid residues. From a practical standpoint, this result suggests that complex peptide samples are best analyzed with both techniques. This strategy is currently being followed by many laboratories and, with regard to interface technology and software support, implemented by the companies that sell both ESI and MALDI mass spectrometers.

Large-scale protein identification by 2D-LC-MS/MS is also referred to as "multidimensional protein identification technology" (MudPIT) [73, 74, 220]. With this approach, protein identification relies solely on MS/MS data. It has been shown, however, that including the LC retention time as peptide-specific information can improve the quality of protein identification significantly, as do very accurate determinations of the molecular mass of the precursor ions (e.g., 1 ppm) [105, 118, 119]. The separation power of this approach, and thereby the number of identified peptides, can be further extended if the sample is separated by SDS-PAGE before proteolysis. The resulting gel lane is cut into several fractions separated by size (e.g., 20 fractions), each of which is proteolyzed in situ and the extracted peptide mixtures are all analyzed by 1D- or even 2D-LC-MS/MS (Fig. 13). An important advantage of prefractionating the sample by size prior to proteolysis is that this information, estimated by comparison with a protein size standard run next to the sample, can be used to confirm protein identifications or reject false positive results. In fact, the gel data provide solid evidence that the peptides identified were indeed part of a protein or a large fragment of it, and not of a small degradation product or already present in the original sample.

Although the combination of SDS-PAGE with 2D-LC-MS/MS is probably the most powerful analytical strategy currently available for the analysis of complex protein mixtures, especially if combined with both ionization techniques of ESI and MALDI, one has to keep in mind that the associated workload and the volume of the raw data, which can easily exceed a terabyte, are enormous and that the interpretation of the results is a challenge. Another disadvantage is that the proteins are only separated by size and that many of them comigrate during SDS-PAGE, which can make it difficult or impossible to differentiate protein species that share large sequence components (domains) or that differ only by secondary modifications or in the exchange or removal of a few amino acid residues. This certainly accounts for the human histones for which, apart from a broad variety of PTMs, many sequence variants encoded by separate genes have also been found that differ only in one or two amino acid residues.

The above approaches are often referred to as "shotgun proteomics", and it has been shown that this strategy has the potential to catalog thousands of proteolytic peptides and assign them to known protein sequences [7, 9, 13, 220, 255]. To judge the challenge of shotgun proteomics it is useful to consider that, on average, a protein cleaved with trypsin will yield 30–50 different peptides and that the digest of complex protein extracts can easily result in a mixture of hundreds of thousands of different peptides, not taking into consideration the many possible posttranslational and other secondary modifications that alter the primary structure of proteins [7]. A second problem that needs to be dealt with is the variation in protein concentration across the sample, which for total DNA-binding protein extracts can easily cover more than six orders of magnitude. Their complete coverage by shotgun proteomics, therefore, is still a wish far away from any experimental reality. If restricted to well-defined subproteomes, however, e.g., all the proteins contained in or associated with the nucleoli, the results that have been published so far are no doubt very impressive (Sect. 5.1).

4.8
Combining MS with Affinity Purification Techniques

Although most chromatographic methods, including cation exchange and RP, in the broadest sense can be considered as affinity-based purification techniques, this term usually refers to far more specific interactions, examples of which are described and discussed in the following. DNA-binding proteins lend themselves to the use of DNA molecules as affinity bait (bait DNA) for their purification and, in fact, this approach seems to be the most promising for systematic identification of those low abundant proteins that interact in a highly specific way with the many different regulatory sequence elements of genomic DNA [256–262].

For affinity purification of proteins that bind to a specific DNA sequence element, this is immobilized on a stationary phase as part of a PCR product or a double-stranded (ds) oligonucleotide. For immobilization, a broad variety of functional groups are available of which biotin or a secondary amine are the most popular, usually attached to the 5′-end of one of the two oligonucleotides used for generation of the capture probe. The fact that custom synthesis and purification of long oligonucleotides (up to 80 bases) including the above functional groups is well established and available in large quantities (> 1 µmol) at an affordable price, makes the use of ds oligonucleotides for affinity purification especially attractive. As stationary phase, magnetic beads are the most popular due to their flexible handling, ease of use, and the possibility of recovering purified molecules in only a few microliters of elution buffer. After affinity purification, the enriched proteins are most often first separated by SDS-PAGE and then identified by MS as described (Sects. 4.1 and 4.7). Alternatively, they are proteolyzed right away and that mixture is analyzed by LC-ESI or LC-MALDI MS and MS/MS.

It is important to notice that if several different proteins were affinity purified by the bait DNA, without additional information it is not possible to distinguish protein–DNA from protein–protein interactions. In other words, in the case of a protein complex, e.g., a transcription initiation complex, only one or two proteins may bind to the bait DNA whereas others interact with them and still others interact with these. Such secondary and tertiary interactions cannot per se be excluded but their likelihood can be reduced by raising the salt concentration in the binding and washing buffers as high as possible without losing the primary DNA-binding proteins. Because DNA–protein complexes usually include a substantial number of tight salt bridges, their interactions can often withstand considerable salt concentrations (several hundred mM NaCl), whereas the above protein–protein interactions are often not stable under these conditions.

In practice, high salt concentrations are one of many necessary means (see below) to reduce nonspecific DNA–protein interactions (protein background). Considering that the difference in free binding enthalpy between

DNA sequence-specific and nonspecific protein–DNA interactions can be very small, it is clear that at the same time when the stringency of the binding and washing conditions is optimized to reduce nonspecific protein background, associated specific protein–protein interactions are at high risk of loss. For this reason, with some exceptions [242–245], DNA-affinity purification has mostly been used for the identification and characterization of DNA-binding proteins and not for protein complexes associated with DNA. This, however, might change in the near future by the systematic use of stable isotopes for differentiating specifically versus nonspecifically bound proteins (Sect. 4.10).

Affinity purification can be extremely powerful (> 10 000-fold enrichment) if high binding affinity can be combined with high binding specificity. Apart from very low copy numbers per cell, this demand can be a major problem when DNA is the bait for fishing proteins since many DNA-binding proteins do not fulfill this criterion in vitro, because either their affinity is too weak or their specificity is poor under these conditions. In practical terms, the consequence is that if the washing conditions applied are too stringent, specific interactants are lost and if they are too mild, they are buried by an overwhelming background of nonspecifically bound proteins.

There are many ways and strategies to deal with this problem. The most important is the use of good controls, e.g., all conditions are the same except the sequence of the immobilized DNA probe. This enables differential display of the results, for example by SDS-PAGE. In this case all bands that are only observed in the sample are interesting candidates, which are then identified and characterized by MS as described. This approach has frequently been applied with varying success [239–241, 263–265]. In some cases it was also possible to differentially display the results directly by MALDI-TOF MS [239], which is a lot quicker than by SDS-PAGE and can yield additional valuable information, especially accurate protein molecular mass data and resolution of protein species close in mass. Direct differential display of DNA-binding proteins has also been demonstrated with SELDI MS [85, 86]. In this case, the DNA was directly immobilized on the active surface area of the SELDI chip. That area, however, is relatively small and, thus, restricts the binding capacity and thereby also the detection sensitivity, especially if the binding affinity is not very high.

Another efficient means to improve the detection sensitivity in DNA-affinity capture experiments is competition, which is easy to implement if sequence-specific DNA-binding proteins are the targets. For this purpose, a large excess of competitor DNA such as Poly-d(IC), which does not contain the bait sequence, is included in the binding as well as the first one or two washing solutions [259, 260]. The competitor molecules are supposed to bind to all proteins that have a high affinity for DNA in general, of which there are many. The problem often encountered is that the introduced competition is insufficient to eliminate substantial nonspecific binding, even if the concentration of the competitor DNA is raised close to saturation [241]. This is a direct consequence of the need for high total protein concentration (most often a crude nuclear extract), which limits the amounts of competitor DNA that can be added without risking precipitation of protein–DNA agglomerates. A second limitation arises from the immobilization of the bait DNA molecules, which reduces their mobility and in return restricts the mobility of bound proteins, resulting in stronger binding compared to the unbound competitor DNA.

An additional means to reduce nonspecific protein background is to preincubate the sample solution one or several times with large amounts of immobilized control DNA [241]. An obvious disadvantage of this strategy is the risk that along with nonspecific DNA-binding proteins, a substantial fraction of the target proteins can also be lost. Nevertheless, this modified approach along with a bundle of other means enabled identification of low abundant transcription factors in human cells (Sect. 5.1) by SDS-PAGE-MALDI MS, which before could not be detected by differential display using SDS-PAGE [241]. That paper is probably the best original reference currently available for a detailed explanation of the problems discussed here and how they can be counteracted. To reduce the total protein mass and increase the transcription factor concentration before DNA-affinity purification, nuclear extracts were first fractionated using phosphocellulose (P11) as stationary phase and an ascending salt gradient (0.1–0.85 M NaCl) for elution. To reduce nonspecific binding of proteins that specifically bind to DNA nicks and ends, instead of Poly-d(IC) short oligo-d(IC) was used as competitor DNA. In addition, both the bait and control oligonucleotides were concatamerized by self-priming PCR before immobilization, with the consequence that considerably more copies of the oligonucleotide sequence than free ends were available for protein binding. Apart from all these means, to identify low abundant human transcription factors it was necessary to tailor protocol details individually, especially the number of preincubation steps, which is acceptable if the scale of the project falls around a few dozen genomic sequence elements. The attractiveness of this approach (succeeding by tuning while doing), however, certainly loses taste when systematic large-scale exploration of gene regulation is the project's goal.

The problems summarized above are not an issue when the target protein is abundant and its affinity for DNA high. In such cases, it is possible to isolate the target proteins in one step and identify and characterize them directly by MS without previous fractionation by SDS-PAGE [239]. However, hunting for low abundant transcription factors or transient interactions can be a daunting task with no guarantee of success. This situation, however, might change in the near future when new MS methods based on stable isotope labeling become an attractive alternative for detecting specifically bound proteins. This will be discussed further below in the context of protein quantification and labeling (Sect. 4.10).

DNA-binding proteins and their complexes can also be enriched by immunoaffinity chromatography prior to MS. A prerequisite is an immobilized antibody that has high binding specificity for the target protein (antigen). A fundamental difference between DNA-affinity purification and all other affinity purification techniques described further below is that the former enables identification of novel DNA-binding proteins, whereas the latter center around known DNA-binding proteins and aim at the identification of protein–protein interactions or PTMs. The success of immunoaffinity purification depends on the quality of the antibody (specificity and binding strength) as well as the expression level of the target proteins. If protein complexes are to be identified, the success also greatly depends on how strong the involved interactions are.

Optimization concerns purity versus loss, and controls should include other antibodies, immobilized the same way but not directed toward any of the target proteins. In immunoaffinity chromatography, another strategic question is whether to use a polyclonal or monoclonal antibody if both are available. In general, this is a decision of higher detection sensitivity versus higher binding specificity. Compared to monoclonal antibodies directed against one specific epitope, polyclonal antibodies represent a family of monoclonal antibodies, each of which is directed toward a different epitope of the same protein. For immunoaffinity purification of protein complexes, this means that the likelihood that the protein escapes antibody recognition because of shielding effects is much higher for a monoclonal than a polyclonal antibody, which suggests the latter class to be the better choice. A disadvantage of polyclonal antibodies is an increased risk of cross reactivity, which should be considered by additional control experiments. Cross reactivity, caused by the antibody background of the organisms in which the polyclonal antibody was expressed (after immunization), to some degree can be differentially displayed by using preimmune serum as control. The benefit of such controls, however, is limited because immune and preimmune serum will not necessarily elicit the same type of background binding. In fact, it is not easy to establish good controls for polyclonal immunoaffinity purification.

Instead of raising antibodies against specific DNA-binding proteins, an alternative approach is to genetically fuse the latter in-frame with an N- or C-terminal peptide sequence tag (epitope tag), which is recognized by a reliable antibody [266]. An important advantage of expressing epitope-tagged recombinant proteins in cell cultures is that one and the same antibody, which has proven efficient for affinity purification, can be used to affinity purify large numbers of proteins and protein complexes, enabling systematic large-scale experiments directed at the identification of protein–protein interaction networks (see below). Fusing candidate genes with a specific gene (or parts of it) to add specific functions to their expression products is a very powerful genetic technique that has found broad application in protein research.

Over the past few years, MS has been combined with many different epitope tags [267–269] as well as other protein or peptide tags that can easily be attached to a recombinant protein by extending or fusing coding DNA. A prominent example for epitope tagging is the flag tag, which has been used, e.g., in a large-scale experiment to identify protein–protein interaction networks in yeast [270]. Describing all the tags that have been used and discussing their advantages and disadvantages falls outside the frame of this contribution. However, one specific technique, called tandem affinity purification (TAP), has proven especially powerful for the systematic study of protein–protein interactions by MS (TAP-MS) [271], and is therefore described in more detail in the following.

The key idea of TAP is to perform affinity purification in two sequential steps using a tandem affinity tag that comprises two different individual tags separated by a rare protease cleavage site. In the original protocol, the tandem affinity tag consists of a protein A tag (distal tag) and a calmodulin-binding peptide (proximal tag), separated by a tobacco etch virus (TEV) protease cleavage site [271] (Fig. 14). During affinity purification, the tagged proteins and protein complexes are first retained and gently washed on IgG-sepharose via the strong affinity of the protein A moiety for IgG molecules. The stationary phase is then incubated with TEV protease, specifically releasing only tagged proteins and their binding partners. The nonspecifically bound protein background is left behind. In the second step, the tagged proteins and protein complexes are retained on calmodulin-sepharose in the presence of calcium ions and, after washing, released by calcium chelation. Again, all nonspecifically bound protein background is left behind. The proteins recovered with the sample and all controls, including the TAP tag alone, are then differentially displayed by SDS-PAGE and all sample-specific bands are identified by ESI or MALDI MS as described.

An important aspect of TAP is that stringent washing with the unavoidable risk of sample loss is replaced by two independent washes on separate supports, allowing for mild conditions without losing washing efficacy. Another equally important aspect is that the recovery conditions in both cases are very gentle, minimizing carryover of nonspecifically bound proteins as well as loss of less tightly bound interaction partners. The superior performance of TAP, compared to a one-step affinity purification using either tag alone, was convincingly demonstrated in the first publication of the technique [271]. The above TAP-MS technique was developed in *S. cerevisiae* and later adapted and applied to other organisms including *Schizosaccharomyces pombe* [272, 273], plants [274, 275], and mammals (see below). Along with the application to different species, new tag combinations also emerged.

The application of TAP-MS to the identification of mammalian protein interaction networks is not nearly as straightforward as it is for yeast for several reasons. First of all, homologous recombination and expression of the tagged proteins under the control of endogenous promoters is difficult in

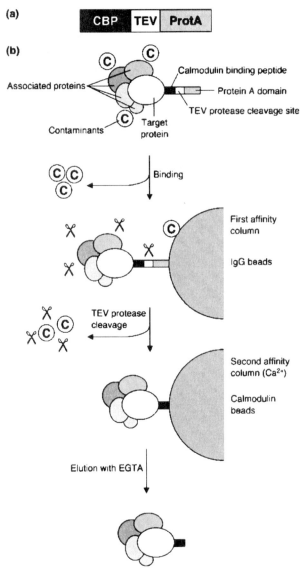

Fig. 14 The tandem affinity purification (TAP) technique. **a** Schematic representation of the TAP tag. **b** Overview of the TAP purification strategy. Reproduced from [10] with permission from © Birkhäuser Verlag, 2006

mammalian cells and not yet feasible on a large scale. Alternative approaches based on expression of cDNA constructs [276–280] suffer from nonphysiological expression levels and competition with the endogenous untagged proteins. The latter limitation has been addressed by using RNA interference (RNAi) to silence the endogenous gene, e.g., TAP-tagged mammalian proteins were expressed in *Drosophila* cells and expression of the homologous endogenous genes was interrupted by RNAi. Other limitations of the TAP technique can arise from the size of the tag (~ 22 kDa for the original TAP tag), which can interfere with protein function and weaken or exclude specific protein–protein interactions. Nevertheless, with regard to the quality and amount of data of protein–protein interactions of DNA-binding proteins generated over recent years (Sect. 5.3), TAP-MS is unsurpassed by any other method available today and, although inherently more difficult, the number of mammalian protein–protein interactions identified by this combination is growing rapidly.

4.9
Combining MS with Biomolecular Interaction Analysis

Biomolecular interaction analysis (BIA) based on surface plasmon resonance (SPR) is a popular affinity-based biosensor technology optimized for monitoring interactions between biomolecules in real time. BIA is an important analytical technique for the study of DNA-binding proteins and, therefore, the topic of a separate chapter of this book. This chapter discusses why and how BIA has been combined with MS, and the benefits and problems of this analytical union.

For proteins with molecular masses up to 50 kDa, the numbers of molecules that can be bound to a BIA chip are the same order of magnitude as those typically needed for ESI or MALDI, indicating that the coupling of BIA with MS is possible. However, what has to be considered is that the sensitivity of BIA increases steadily with increasing molecular mass of the bound molecules, whereas the detection sensitivity of MS decreases with increasing mass. One strategy for combining BIA with MS (BIA-MS) is to use the sensor chips as MALDI sample support [281–284]. In this approach, binding and washing of the target proteins is monitored and quantified by SPR. Then the sensor chip is taken out of the instrument, matrix solution is applied to the active surface, and the bound proteins are analyzed by MALDI-TOF MS. A second strategy avoids destruction of the sensor chip and, instead, recovers the bound analyte molecules by elution and submits the eluate for identification to ESI [285] or MALDI MS [286]. This approach has the advantage that the expensive sensor chips can be reused many times, and the eluted proteins, besides determining their molecular mass, can also be identified and further characterized on the level of proteolytic peptides (Sects. 4.1–4.3).

An enormous advantage of BIA based on SPR, compared to MS, is label-free absolute quantification of protein interactions in real time. Qualitative information about the identity of the trapped molecules, however, solely relies on the affinity of the immobilized ligands. For instance, if oligonucleotide probes are immobilized to affinity-capture specific DNA-binding proteins out of a solution, BIA will report this event over time as a change of total protein mass that is held in place by the immobilized DNA. If the solution contains only one protein species and its molecular mass is known, this information can be directly translated to the number of affinity-bound proteins over time, which provides the basis to determine kinetic rate constants.

If the solution, however, contains many different proteins and it is not clear which of them interact with the immobilized DNA under the experimental conditions applied, the raw data of the above experiment are of little use. The missing information is the identity of the affinity-bound proteins. Generating this information is the strength of ESI and MALDI MS, which explains the interest in combining the two analytical techniques [282, 287–292]. If, however, more than one protein species is affinity-purified, which is very likely if the above protein mixture is a crude nuclear extract, the situation is more complicated. In this case, in addition to qualitative data (protein identities), information about the relative amounts of each of the affinity-purified protein species is required to calculate absolute numbers. How ESI and MALDI MS can be used to generate this information is described in the following section.

4.10
Protein Quantification and Stable Isotope Labeling

One very important analytical aspect in the study of proteins in general and DNA-binding proteins in particular is quantification. Measurements of their concentration over time, determined by their rate of synthesis, modification, and degradation, are essential to identify and understand their many different functions and interactions. Furthermore, new research approaches such as systems biology rely on quantitative protein data as the input for modeling biological processes. Consequently, as proteomics turns quantitative and MS is a key technology for protein analysis, MS-based quantification methods and strategies are strongly sought after.

Quantitative data exist in two forms: absolute or relative numbers. For instance, in the process of drug development and validation, accurate determination of absolute compound concentrations is a daily requirement, which for protein-based candidates can be hard to get by. On the other hand, for the understanding of biological processes relative differences are often more informative than absolute numbers. Absolute quantification of protein concentrations in all the samples that are to be compared, of course, renders relative measurement unnecessary as these data can be calculated from the former in any relation. The opposite direction, to determine absolute amounts by relative measurements, requires a reference compound of known concentration, which can be added to the samples to be analyzed. This is common practice and in fact most absolute numbers are based on relative measurements.

For protein researchers, the main limitation for absolute quantification arises from the lack of suitable reference proteins or peptides. Other problems are connected to their storage and handling as well as their costs. In contrast, relative measurements can be conducted far more easily and be scaled up to large numbers of proteins for a fraction of the effort and costs of absolute data, as will be explained below.

For mass spectrometric quantification, signal intensities or integrals are related to each other. The result is relative numbers (fold changes) that turn into absolute values at the moment the concentration or amount of the reference compound in the sample is known. As any difference in the chemical structure of analyte and reference molecules can affect losses during the sample preparation as well as ESI or MALDI, they should ideally only differ in mass but not in their chemistry. This can be accomplished by the exchange of ^{12}C, ^{14}N, or ^{16}O atoms by their stable isotopes ^{13}C, ^{15}N, and ^{18}O. The result is two compounds that only differ in the number of neutrons they contain, but are otherwise chemically equivalent. The exchange of hydrogen atoms by deuterium, which is easier (cheaper) to realize, is a less optimal choice because these two isotopes differ chemically significantly, e.g., RP LC can separate differently deuterated peptides.

The concept of stable isotope incorporation is at the core of all accurate MS-based protein quantification techniques and can be applied on the levels of intact proteins as well as their proteolytic peptides [37, 293–298]. In the latter case, which is the most popular approach, three or four additional neutrons are sufficient to distinguish the signals of labeled and unlabeled peptide. To minimize the risk for errors caused by overlapping isotopic distributions, however, six additional neutrons is a better choice. For peptide molecular ions exceeding m/z 5000, that number should be increased accordingly. For MS-based protein quantification, an important advantage of analyzing proteolytic peptides, besides higher detection sensitivity and signal resolution, is the possibility to improve statistics, and thereby reduce the error, by comparing the signal intensities of several pairs of labeled and unlabeled peptides.

Many analytical strategies have been developed for stable-isotope-based quantification of proteins, which all differ in the way stable isotopes are introduced into peptides or proteins (Fig. 15). They can be classified as: (1) metabolic labeling, where cells acquire stable isotopes from the growth medium and incorporate them during protein biosynthesis;, (2) enzymatic labeling, where stable isotopes are incorporated by an enzymatic reaction performed in vitro; (3) chemical labeling, where stable isotopes are introduced by a chemical reaction in vitro; and (4) spiking in a labeled reference compound.

One straightforward approach for the introduction of stable-isotope-labeled peptides is to chemically synthesize them and add known quantities to the sample [299]. It extends to peptides with the well-established technique of stable isotope dilution, which is routinely used in pharmaceutical research for the quantification of small-molecule-based drugs. This approach is the most powerful and the most expensive available for absolute peptide

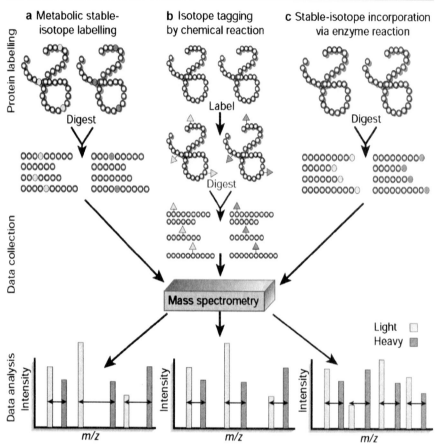

Fig. 15 Schematic representation of methods for stable-isotope protein labeling for quantitative proteomics. **a** Proteins are labeled metabolically by culturing cells in media that are isotopically enriched (e.g., containing ^{15}N salts or ^{13}C-labeled amino acids) or isotopically depleted. **b** Proteins are labeled at specific sites with isotopically encoded reagents. The reagents can also contain affinity tags, allowing for the selective isolation of the labeled peptides after protein digestion. The use of chemistries of different specificity enables selective tagging of classes of proteins containing specific functional groups. **c** Proteins are isotopically tagged by means of enzyme-catalyzed incorporation of ^{18}O from $H_2{}^{18}O$ during proteolysis. Each peptide generated by the enzymatic reaction carried out in heavy water is labeled at the C-terminus. In each case, labeled proteins or peptides are combined, separated, and analyzed by MS and/or MS/MS to identify them and determine their relative abundance. The patterns of isotopic mass differences generated by each method are indicated schematically. The mass difference of peptide pairs generated by metabolic labeling is dependent on the amino acid composition of the peptide and is therefore variable. The mass difference generated by enzymatic ^{18}O incorporation is either 4 or 2 Da. The mass difference generated by chemical tagging is one or multiple times the mass difference encoded in the reagent used. Reproduced from [7] with permission from © Nature Publishing Group, 2006

quantification. That, however, might change when the interest in the technique increases further. Custom synthesis of peptides in large numbers and quantities is not an issue anymore, and the prices for stable isotope incorporation will certainly drop when their popularity grows further. Alternatives are enzymatic or chemical labeling after peptide synthesis (see below).

For absolute protein quantification, limitations on the use of stable-isotope-labeled peptides arise from the late state of sample preparation where they can be added, i.e., during or after proteolysis. Conclusions back to the amount of protein that was in the original sample are therefore limited. The obvious way around this problem, that is, using chemically synthesized isotope-labeled proteins as reference compounds, except for some small species, is currently not an option due to the difficulties in correctly synthesizing, folding, and modifying long amino acid chains. If the target protein, however, is available as a recombinant expression product, metabolic labeling by the host organism can be an option worth pursuing, especially if large-scale absolute quantification experiments are planned.

Metabolic labeling of peptides and proteins with stable isotopes is a well-established technique that has been used for decades to assist their structure analysis by NMR [297]. This is achieved by growing the host cells in a medium highly enriched in the stable isotopes to be incorporated. For simple organisms like bacteria and yeast, this can be realized in a cost-effective way, e.g., by providing ($^{15}NH_4$)$_2SO_4$ as the only available nitrogen source, and these microorganisms can in turn be fed to small organisms such as *Caenorhabditis elegans* or *Drosophila melanogaster* [300]. For higher developed eukaryotic cells, which have lost the ability to synthesize all amino acids themselves, the necessary substitution in the growth medium is more complex and more expensive.

Metabolic stable isotope labeling is an efficient technique because all proteins of the cells involved are labeled [297]. This renders relative quantification experiments that include many proteins or even complete proteomes straightforward. There are two ways for conducting such experiments; which of them is the better choice depends on the analytical question to be answered. One is to compare different cell cultures grown in parallel and the other is to compare aliquots of the same culture taken at different time points. To compare the proteome (or a subset of it) of two cell cultures, e.g., one opposed to a drug and the other not, one is grown in stable-isotope-enriched medium and the other in normal medium. At a defined point, equal amounts of protein extracts from the two cultures are mixed, separated, and analyzed by MS as described. Because the labels do not interfere with the biological processes involved, nor do they affect the sample preparation, any available technique for protein separation and further processing can be applied.

To compare different aliquots derived from the same culture, e.g., before and after onset of a stress regime, at a defined point in time the growth medium is replaced by one of different isotope composition. The consequence is that past this exchange, with a certain delay, all new proteins that are synthesized differ in their isotope composition from those that were synthesized before. Thus, proteins synthesized after medium replacement can be quantitatively distinguished from those already existing. Monitoring that relation over time enables determination of protein turnover rates.

It has also been demonstrated that this approach is even more powerful if combined with an additional means to compare protein species concentrations across the different extracts (time points), independent of when they were synthesized [301]. This was realized by separating the different extracts by 2-DE, comparing protein spot staining intensities across the different gels, identifying the contained protein species by MALDI MS PMF, and determining for each spot on each gel the ratio of labeled versus unlabeled protein by comparing the signal intensities of labeled and unlabeled peptides in the recorded PMF spectra. By correlating changes of the total concentration of individual protein species over time, estimated by 2-DE, with the corresponding ratios for labeled versus unlabeled peptides determined by MS, it was possible to distinguish the influence of synthesis rates versus degradation rates on the present concentration of many different protein species, which is important for understanding and modeling the regulation of protein expression levels.

Stable isotope labeling by amino acids in cell culture (SILAC) [297, 302] is another powerful approach of metabolic labeling. Amino acids containing stable isotopes, such as arginine bearing six ^{13}C atoms, are supplied for growth resulting in their incorporation into newly synthesized proteins in dependence on their sequence. There are many different ways and strategies to perform SILAC experiments, and different amino acids and combinations of them can be used to address specific questions [37, 112, 297, 302–307]. For bacteria and yeast, to avoid interference with in vivo synthesis of the supplied amino acids, auxotrophic strains such as Arg$^-$/Lys$^-$ double auxotrophic yeast [112] are available, which eliminate this problem. Another possible interference in some cell types is metabolic conversion of labeled arginine to labeled proline, which is not intended. This problem can be minimized by titrating the amount of labeled arginine added to the growth medium [37].

A special strength of SILAC is the ease of multiplexing by supplying differently labeled amino acids to different cell cultures or at different time points. For instance, three different forms of arginine have been used to compare phosphotyrosine proteins at five time points of epidermal growth factor (EGF) stimulation [149]. SILAC also enables quantification of posttranslational modification events [308], e.g., protein methylation, by supplying labeled methionine, which as part of S-adenosylmethionine is the primary methylation donor in biological systems [307]. Culture-derived isotope tags (CDIT) further extend the application of metabolic labeling by using SILAC labeled cells as the bridging internal standard between two tissue samples [309]. This approach was shown to enable quantification of several hundred protein species, extracted from mouse brain sample by using labeled proteins of a Neuro2A cell line as internal standard.

Natural limitations of metabolic labeling for MS-based protein identification arise from the biological level where it works best, i.e., cell lines (and not tissues), as well as the level of isotope enrichment that is achievable in adequate time. The fact that the mass difference between labeled and unlabeled proteolytic peptides depends on their amino acid sequence can be considered as an advantage or a disadvantage. For instance, when labeled with ^{15}N, the mass difference determined for a proteolytic peptide reports the number of nitrogen atoms it contains, which is specific information that can be used to confirm or reject PMF or MS/MS-based identification results. However, the fact that the assignment of corresponding peptide signal pairs in the recorded mass spectrum requires peptide-specific knowledge can also be considered as a limitation. Systematic screening for signal pairs representing labeled and unlabeled peptides is certainly more straightforward if their mass difference is a known constant or a multiple of it. This can be achieved by SILAC as well as chemical or enzymatic labeling described and discussed in the following.

The most prominent method for enzymatic labeling is based on the exchange of one or two C-terminal carboxyl ^{16}O atoms by ^{18}O in the presence of $H_2^{18}O$ catalyzed by an endoprotease, most often trypsin, during or post protein proteolysis [296, 298, 310–316]. Other endoproteases such as Lys-N are less well suited for this task, because they only catalyze the exchange of one oxygen atom. With trypsin, the exchange reaction is not always complete and to drive it near 100% can require up to 48 h incubation time, which can be considered a disadvantage. Furthermore, even if the exchange is 100% the resulting mass difference of 4 Da is still too small to avoid overlap of the isotopic distributions of larger tryptic peptides. However, it has been shown that if this problem is taken care of by baseline resolution, proper ion statistics, and software that corrects the intensity of overlapping signals, accurate quantification with relative errors not greater than a few percent is possible. A major advantage of enzymatic (and chemical) labeling is that it is independent of protein synthesis, and can therefore be applied to any biological sample including human tissue sections. This is what makes the above approach so attractive and has led to numerous applications and modified versions of the original protocol.

Chemical labeling of peptides and proteins has become a very broad field that covers many different methods and strategies [22, 37, 116, 296–298, 317–352], and not a month passes by without a new technique or modified version of an existing protocol being published. What makes the use of chemical reagents (tags) so popular is the possibility to optimize (design) them for specific applications, e.g., to label exclusively specific amino acid residues or PTMs [308], and to include other functional groups (multifunctional labels) that aid the subsequent sample preparation or mass spectrometric analysis [353].

The best known example of such a multifunctional reagent is probably the isotope-coded affinity tag (ICAT) described in 1999 [317]. It consists of a cysteine-directed reactive group, an oligoether linker region harboring eight deuterons ("heavy" label) or hydrogen atoms ("light" label), and a biotin group for affinity purification (Fig. 16). Cysteine was chosen as target because it is less frequent than most other amino acids, and therefore allows efficient reduction of the complexity of total proteolytic digests of complex protein samples, which simplifies "shotgun proteomics" applications. The ICAT was the first isotope-labeled reagent commercialized for MS-based quantification of proteins and has been successfully applied but also heavily criticized for its drawbacks. The problems encountered ranged from undesired side reactions, its restriction to cysteine residues, which not all proteins contain, negative interference with identification of the labeled peptides by MS/MS, and the use of deuterium as "heavy" isotope, which is known to interfere with RP LC (see above). This led to further developments, which made ICAT more practical by using a cleavable and coeluting tag [318–320].

A description of all reagents and strategies that have been explored for MS-based protein quantification would require a book on its own, and therefore cannot be provided here. However, what can be stated is that for good protein coverage, – COOH and – NH$_2$ groups are the most popular and best-suited targets for labeling and can, in principle, be used to quantify any peptide and protein. A typical example for modifying carboxyl groups is permethylation esterification. For primary amine groups (lysine and amino terminus), very efficient and specific reactions, such as succinylation using nicotinoyloxy succinimide as reagent [324], are available [325–327]. The isotope-coded protein tag (ICPL) is a recently commercialized variant that has been optimized for the labeling of complex protein samples before proteolysis.

An innovative technique that has also recently been commercialized and has gained popularity in a short time takes the application of chemical incorporation of stable isotopes for protein quantification a step further, i.e., quantification is based on MS/MS and not MS data. It utilizes an isobaric tag for relative and absolute quantification (iTRAQ), which targets primary amine groups and releases a specific reporter ion when the peptide molecular ions are fragmented. There are four isotope-coded variants of the tag that result in the release of four different reporter ions of mass 114, 115, 116, and 117 Da. Before fragmentation, the mass of each of the four reporter groups is balanced by a carbonyl group, also part of the tag, such that the total mass of the tag is always the same. As a consequence, four aliquots of the same peptide labeled with the four different tags are isobaric (have the same molecular mass).

An important consequence is that the complexity of the mass spectra acquired from mixed samples is not affected by the tag (no pairs of signals) and the detection sensitivity is not compromised, as it is if the total signal intensity observed for each peptide is shared by a "heavy" and a "light" variant of

Analysis of DNA-Binding Proteins by Mass Spectrometry

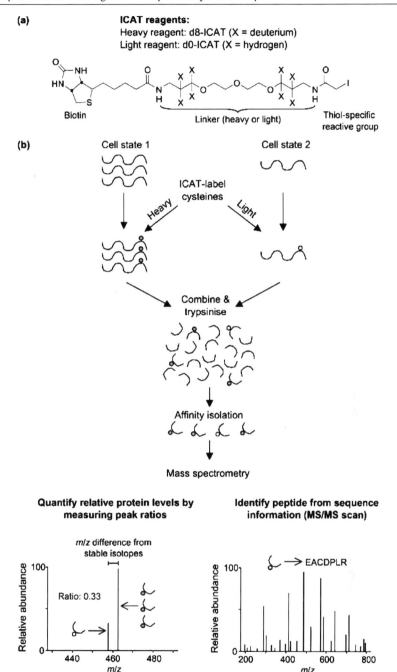

Fig. 16 The ICAT strategy for quantifying differential protein expression. **a** Structure of the ICAT reagent. **b** Schematic of the ICAT strategy. Reproduced from [10] with permission from © Birkhäuser Verlag, 2006

it. Quantification is based on MS/MS spectra where the signal intensities of the different reporter ions are correlated. According to its inventors and some independent reports, the iTRAQ also enhances fragmentation yields and does not compromise peptide identification by MS/MS. An important feature of the iTRAQ, which makes it especially attractive for relative protein quantification, is that four separately labeled samples can be pooled and together separated, identified, and then quantified relative to each other.

The applications and techniques of stable isotope incorporation go far beyond MS-based quantification [298]. An emerging new application that has gained significant attention among scientists is the use of stable isotopes in MS-based interaction proteomics. As described for the affinity purification of DNA-binding proteins, these experiments suffer from the inevitable trade-off between detection sensitivity and binding specificity. Stable isotope labeling now promises a way out of this dilemma [320, 354, 355], by providing an efficient means to differentiate specific versus nonspecific interactions by MS. This is made possible by differentially stable isotope labeling the proteins that are enriched by the sample (bait) and the control. This can be done before or after affinity enrichment. Either way, the two fractions (bait and control) are afterward mixed 1 : 1. If the control is close to the sample, e.g., everything is the same except for the sequence of the immobilized DNA, proteins that bind nonspecifically to DNA, or the support material used, are equally abundant in the two pull-down samples whereas specific interactions with the bait or the control DNA result in differential ratios. In the subsequent MS analysis, irrespective of whether the enriched proteins or their proteolytic peptides are looked at, signal pairs of comparable intensity (light and heavy variants) refer to nonspecific interactions and single signals and all signal pairs that differ significantly in their intensity refer to specific interactions. Because specific interactions with the control DNA in general cannot be excluded, the origin of all single signals (from the control or the sample) needs to be determined by MS/MS.

The feasibility of the above concept was demonstrated convincingly [320] and subsequent work supports its value and has extended its applicability [242–245, 354, 355]. Both metabolic and chemical labeling were successfully applied. It is foreseeable that the application of stable isotope labeling for the identification of low abundant or transient protein interactions will undergo further developments and adaptations. This assumption is supported by a recent publication that demonstrates enzymatic labeling to be an attractive alternative and proposes a new strategy, which renders it possible to recognize all peptide molecular ions specific to the sample solely based on the absence of signal doublets in the recorded peptide mass spectra [316] (Fig. 17). This is achieved by splitting the control, labeling one half with ^{18}O and labeling the other half as well as the sample with ^{16}O. Subsequently, the three solutions are mixed in the ratio 2 : 1 : 3, respectively, and analyzed by LC-MS. The advantage of this approach is that all proteolytic peptides present

Analysis of DNA-Binding Proteins by Mass Spectrometry

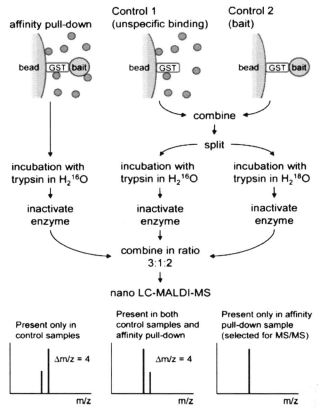

Fig. 17 A new analytical strategy for comparing protein mixtures. Three protein samples, for example, obtained from one affinity pull-down and two control experiments, are compared with the objective of identifying the proteins that are only present in one of them, e.g., the affinity pull-down isolate. For this purpose, the latter is incubated with trypsin in $H_2^{16}O$. The two controls are pooled, then split into two parts, which are incubated separately with trypsin, one part in $H_2^{16}O$ and the other in $H_2^{18}O$. The resulting peptide mixtures are mixed in the ratio 3:1:2 and analyzed by nano-LC-MALDI MS and MS/MS. Tryptic peptides of proteins present only in the control samples or in the control samples and the affinity pull-down isolate are detected as paired signals with a distance of 4 Da, while peptides detected only in the affinity pull-down isolate are detected as unpaired signals. Reproduced from [316] with permission from © American Chemical Society, 2006

only in the sample will be detected as singlets, whereas all others will be detected as doublets, which renders the assignment of peptides specific to the sample and their identification by MS/MS straightforward (Fig. 18). For peptides that are only present in the control, the intensity ratio for the heavy and light variants falls around 2:1, whereas for those which are also present in the sample, it approaches the opposite ratio of 1:2, which renders it possible to distinguish these two situations as well.

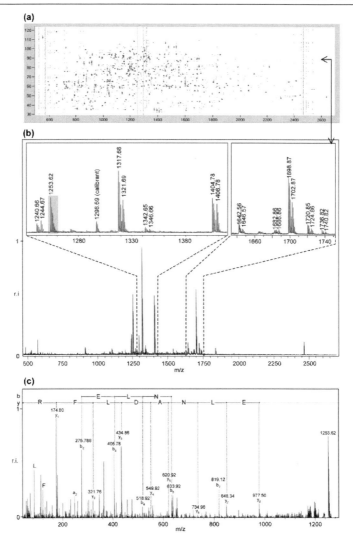

Fig. 18 The new strategy outlined in Fig. 17 was evaluated in the analysis of protein-affinity pull-down experiments, to distinguish specifically enriched sample components from experimental artifacts. The carboxy terminus of Hsp70-interacting protein (CHIP) was expressed as GST fusion protein in *E. coli* and used as bait to affinity purify potential interaction partners from a crude human brain protein extract. **a** Peptide profile obtained from the LC-MALDI MS analysis of the protein isolate obtained from the affinity pull-down experiment. *Blue dots*: unpaired peptides, *red dots*: paired signals, *black dots*: all detected monoisotopic peaks. **b** Mass spectrum acquired from fraction 89. The signal of m/z 1253.62 is, apart from the internal calibrants, the only unpaired signal in the spectrum, indicating that it originates from a protein that specifically interacts with CHIP. **c** MS/MS spectrum of m/z 1253.62. A database search identified the peptide FEEL-NADLFR of heat shock cognate 71-kDa protein. Reproduced from [316] with permission from © American Chemical Society, 2006

To deal with the complexity of the above samples, which can include a large excess of background proteins, the generated proteolytic peptides are usually separated by RP LC before they are analyzed by ESI or MALDI MS. If that is insufficient, the proteins can be separated by SDS-PAGE before they are proteolyzed (Sect. 4.7). The resulting workload can be regarded as the price one has to pay for a significantly increased detection sensitivity made possible by applying less stringent binding and washing conditions. Obviously, if such experiments are expected to become daily bread, investments in automated workflows are worth a thought and certainly a requirement for any large-scale project.

There are more benefits and applications of stable isotope labeling for the identification and characterization of proteins by MS [298]. Examples are identification and sequencing by MS/MS. For instance, if the heavy and light variants are incorporated at one end of a proteolytic peptide, as is the case for the enzymatic incorporation of ^{18}O (C-terminal carboxyl group), the generated fragment ion series of the heavy and light forms of each peptide will differ in this respect. Based on the observed mass shifts between the two fragment ion spectra, the fragment ions are classified as containing the labeled C-terminus or not (N- or C-terminal fragment ion series). This information can then be used in the data interpretation (de novo sequencing) or as additional input information for database searches or afterward, to confirm or reject the retrieved sequence candidates.

The description and discussion of MS-based methods for the identification and characterization of DNA-binding proteins finishes here. What comes next are examples that illustrate their performance and limitations.

5
Applications and Examples

5.1
Identification and Quantification

Over the past few years, many new DNA-binding proteins have been identified by the use of ESI and MALDI MS. This includes predicted gene expression products of hitherto unknown function as well as known proteins that were not known to interact with DNA. As a consequence of this ongoing discovery, the number of proteins known to be involved in the regulation of DNA expression, replication, recombination, and repair has grown considerably. It is now clear that along with chromatin remodeling and methylation, a broad repertoire of secondary histone modifications as well as the specific binding of transcription factors and many other regulatory proteins to cognate DNA sequence elements all constitute crucial steps in transcriptional activation or silencing.

The examples of MS-based identification and quantification of DNA-binding proteins published over recent years can be classified by the scale of the performed experiments. Most of these were directed at the identification of a few proteins which specifically bind to a selected DNA sequence element. At the other end of the line are large-scale experiments directed at the identification of complete organellar proteomes including many different DNA-binding and DNA-associated proteins. The examples discussed below start with the latter category and end with the identification of individual, highly specific DNA-binding proteins.

With respect to the scale of the project and its impact on our molecular biological knowledge, the most spectacular example is probably the identification of the human nucleolar proteome by SDS-PAGE–ESI MS/MS [305, 356–358]. That subproteome was found to comprise at least 692 different proteins, 126 of which were novel, hitherto uncharacterized proteins, and the large majority of the identified proteins were previously unknown to be associated with the nucleolus [305]. Classification by functional categories revealed that the nucleolar proteome harbors at least 15 kinases/phosphatases, 16 chromatin-related factors, 8 proteins involved in DNA repair, 18 participating in DNA replication, 30 transcription factors, 9 RNA polymerases, 64 RNA-modifying enzymes and related proteins, 76 ribosomal proteins, 33 RNA helicases, 38 splicing related factors, 10 chaperones, and 25 cell-cycle proteins. Furthermore, the flux of 489 of the 692 identified nucleolar proteins was quantified over time in response to three different metabolic inhibitors that each affect nucleolar morphology [305, 357] (Fig. 19). It was found that the nucleolar proteome changes significantly over time in response to changes in cellular growth conditions, and that proteins that are stably associated, e.g., RNA polymerase I subunits and small nuclear ribonucleoprotein particle complexes, exit from or accumulate in the nucleolus with similar kinetics, whereas protein components of the large and small ribosomal subunits leave the nucleolus with markedly different kinetics.

In the most successful experiments, the nucleoli were isolated and purified by density gradient centrifugation, the contained proteins fractionated by SDS-PAGE, in-gel digested with trypsin or Lys-C, and the extracted proteolytic peptides of each gel fraction were separated and identified by RP nano-LC-ESI MS/MS [305, 356, 357]. Relative quantification was performed by using the SILAC technique (three different variants of arginine) to label individual cell culture samples (time points). By using one sample as common reference (time point zero) to which in each experiment two others are related, up to nine samples (time points) were analyzed to monitor changes of the nucleolar proteome over time in response to the drug [305, 357].

The above experiments can be considered as a breakthrough that has broadened and changed our picture of the complexity and function of the nucleolus. The first, incomplete list of human nucleolar proteins comprising 272 different proteins, which was published in 2002 [356], has already

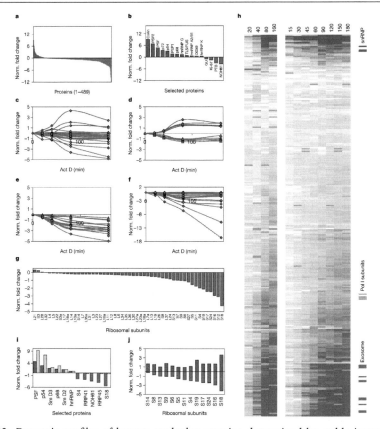

Fig. 19 Dynamic profiles of human nucleolar proteins determined by stable isotope labeling (SILAC) and MS. *Red* indicates recruited proteins and *green* indicates depleted proteins. **a** All proteins showing change from first to last time points. **b** Proteins known from the literature to be recruited to or depleted from nucleoli upon treatment with actinomycin D (Act D). **c** Different kinetic profiles for different DEAD box proteins (*top curve* to *bottom curve*: BAT1, CHD4, DDX10, DDX17, DDX18, DDX21, DDX24, DDX27, DDX31, DDX3X, DDX48, DDX49, DDX5, DDX50, DDX51, DDX52, DDX54, DDX56, DHX33, DHX37, MTR4, RUVBL2). **d** Dynamic profile for polymerase I subunits (*green*; POLR1C, PAF53, POLR1B, POLR1A, POLR1D, TTF1, UBTF) and snRNP proteins (*red*; SNRPB, SNRPA, SNRPD2, SNRPD3, SNRPF). **e** Dynamic profiles for subunits of the exosome (*green*; RRP42, RRP46, RRP40, RRP4, RRP43, RRP41, RRP45, CSL4, RRP44, MTR3, MTR4) and the RNase P (*blue*; RPP14, RPP25, RPP38, RPP30, POP1). **f** Dynamic profile of the human homologs of the yeast SSU processome proteins. **g** Fold change of the large (*blue*) and small (*green*) ribosomal subunit proteins. **h** Hierarchical clustering of 302 proteins using fold change data from five and nine time point experiments. The indicated proteins are snRNP (SNRPA, SNRPD2, SNRPD3), Pol I (POLR1A, POLR1D, POLR1B, POLR1C), and exosome components. **i** Comparison of fold change for a subset of proteins upon treatment with actinomycin D (*red/green*, average fold change after treatment for 80 and 160 min) and DRB (*yellow*, 80 min). **j** Comparison of fold change for the small ribosomal subunit proteins upon treatment with actinomycin D (*green*) and MG132 (*red*, 8 h). Reproduced from [305] with permission from © Nature Publishing Group, 2006

gained significant attention and provided the basis for other, previously impossible research, especially in the emerging field of bioinformatics. A good example is the recent investigation of nucleolus evolution based on sequence analysis techniques that identify known and novel conserved protein domains. Using the sequences of the above 272 proteins as input data, 115 known and 91 novel nucleolar protein domain profiles were identified. Correlating these across a collection of complete proteomes of selected organisms confirmed the archaebacterial origin of the core machinery for ribosome maturation and assembly, but also revealed substantial eubacterial and eukaryotic contributions to the nucleolus proteome and how these affected its evolution [359].

Identification of the protein complement of murine nuclear interchromatin granule clusters (nuclear speckles) [360], yeast nuclear pores [361], the yeast and human spliceosome [362–365], and the human centrosome [366] are further examples of successful organellar proteomics [367] based on large-scale "shotgun" protein identification and quantification using stable isotope labeling. Over the past 10 years, the identification and characterization of multiprotein complexes as well as the individual transcription factors involved in gene expression by ESI and MALDI MS has been, and still is, a hot topic. For detailed information regarding this specific application of MS, the reader is referred to the comprehensive review of Cameron et al. [164]. Another example of the characterization of a multiprotein complex is the yeast RNA polymerase II (Pol II) preinitiation complex. After DNA-affinity purification, 45 of the known Pol II core proteins as well as a novel, tenth subunit of the TFIIH complex were identified by MS, and differential ICAT labeling was used for distinguishing specifically and nonspecifically bound proteins [242, 243].

Other examples are the identification of the proteins NOT1 and NOT3 as components of the yeast CCR4 transcriptional complex by immunoprecipitation followed by SDS-PAGE and MALDI MS PMF [368]. The same strategy was used to identify the transcription factor elf5 as an additional, hitherto unknown component of the elf3 transcription-factor complex [369, 370]. In a separate study, the protein kinase Tra1p was identified as a subunit of the yeast Ada-Spt transcription complex [371]. N-CoR, a human nuclear-receptor corepressor, is known to be a central component of different multiprotein complexes involved in multiple transcriptional processes. MALDI MS analysis of the N-Cor-1 complex purified from HeLa cells yielded a series of hitherto unknown subunits including the SWI-SNT-related proteins BRG1, BAF170, BAF155, FBAF47/INI1, and KAP-1 [372].

An impressive example of the important role of MS in the study of gene expression regulation by transcription factors is a study of the peroxisome proliferator-activated receptor (PPAR) proteins regulating genes involved in mammalian lipid metabolism [373]. In this example, 2-DE-MS was used to study the effect of a therapeutic dose of the PPARα transcription factor in a mouse disease model of insulin resistance and diabetes. The dose caused induction of peroxisomal fatty acid β-oxidation in obese diabetic mice, and a differential analysis of the liver proteome by 2-DE-MS identified 16 proteins that were upregulated in response to the treatment.

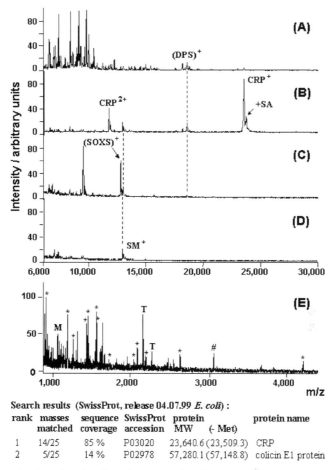

Fig. 20 Identification of cyclic adenosine monophosphate receptor protein (CRP) in *E. coli* crude cell extract by DNA-affinity purification using magnetic particles as solid support. **a** Mass spectrum obtained from 0.5 µL crude cell extract (input) diluted 1 : 50 with MALDI matrix solution. **b** Mass spectrum of proteins isolated with an immobilized DNA probe containing a CRP-binding consensus sequence. **c** Mass spectrum of proteins isolated with a control DNA probe lacking a binding motif for CRP. **d** Mass spectrum of proteins isolated without immobilized DNA. **e** Identification of CRP by peptide mass fingerprinting. *Identified CRP tryptic peptides; #, the first 26 amino acid residues of CRP lacking the N-terminal methionine; + *SA*, adduction of one sinapic acid molecule used as MALDI matrix; *SM*, streptavidin monomer; *M*, matrix signal; *T*, trypsin autolysis products, used for internal spectrum calibration; *DPS*, DNA-binding protein from starved cells. Reproduced from [239] with permission from © Nature Publishing Group, 2006

In the following, a few examples are provided that document the application of DNA-affinity purification combined with MS for the identification of transcription factors and other highly specific DNA-binding proteins. One of the first examples is the identification of the components of the ARF6 transcription factor complex [265]. It was shown that ARF6 is a heterodimeric complex of the two nuclear hormone receptors PPARγ and retinoid X receptor alpha (RXRα). Another example is the identification and characterization of the *E. coli* cyclic adenosine monophosphate (cAMP) receptor protein (CRP) in crude cell lysates [239]. Affinity purification was performed using biotinylated double-stranded oligonucleotides that were immobilized on streptavidin-coated magnetic particles and contained a known consensus sequence to which CRP binds specifically in the presence of cAMP. Control experiments included the absence of cAMP in the binding and washing buffer as well as a nonspecific double-stranded oligonucleotide probe. CRP was identified by PMF, and a direct differential display of the purified samples by MALDI-TOF MS confirmed what was expected, i.e., the purification yield of CRP was much reduced in the absence of cAMP and not detectable with the control DNA (Fig. 20).

The same approach was used to identify rat RXRα expressed in the yeast strain BJ2168, serving as a background-free system. For affinity purification a double-stranded oligonucleotide was used that contained a direct repeat with one base-pair spacing (DR1) from the rat acyl-CoA oxidase promoter, which had previously been shown to direct RXR-dependent transactivation of a reporter gene in yeast. The low abundant human transcription factors AP.1 and PU.1 expressed in promyelocytic leukemia NB4 cells, however, could only be identified when a series of additional means to separate specific from nonspecific DNA-binding proteins was included in the affinity purification protocol including repeated preincubation of the sample with immobilized control DNA (Sect. 4.8) [241].

5.2
Posttranslational Modifications

Identification and characterization of the many different PTMs that DNA-binding proteins can be subject to is important to understand their functions and how these are controlled. The best-known examples of this are certainly the histones, which are subject to nearly all PTMs known for DNA-binding proteins including acetylation, methylation, phosphorylation, ADP-ribosylation, biotinylation, sumoylation, and ubiquitylation. For this reason the histones have been chosen as examples to illustrate the application of ESI and MALDI MS for the identification and characterization of PTMs.

The complex patterns of site-specific PTMs, which were first found within the N-terminal tails of the different histones, led to the postulation of the existence of a histone code (Sects. 1 and 4.3) [138, 139], for which Turner et al.

later introduced a specific nomenclature [141] that allows patterns of histone modification to be clearly and unambiguously specified. These are listed starting from the left with naming of the histone affected (e.g., H2B), then the amino acid residue and its location (e.g., R17) and the corresponding modification, followed by the next residue and its modification and so on until all modifications are listed. For all known modifications a list of unambiguous abbreviations was suggested, which makes it possible to specify each of them with a few symbols. For instance, ac, me, ph, and ub stand for acetylation, methylation, phosphorylation, and ubiquitylation, respectively. If more than one of these groups can be attached to the target residue (e.g., mono-, di-, or trimethyl lysine), this is indicated by a number following the abbreviation (e.g., H3K4me1, H3K4me2, and H3K4me3). In the case of dimethyl arginine the two known structures, i.e., each of the terminal nitrogen atoms carries one methyl group or one carries both, are specified by the letter s (symmetrical) or a (asymmetrical) (e.g., H3R2me2s and H3R17me2a).

It is interesting to note that the original assumption that PTMs of the core histones are mostly directed toward their flexible N-terminal tails turned out to be wrong as soon as MALDI and ESI MS instead of Edman sequencing were applied for histone modification analysis, reflecting the analytical bias of the latter technique. It was also found that the C-terminus, as well as the entire central region of the core histones, is subject to PTMs [374–376]. An impressive example of the impact of MS on the study of histone PTMs is the identification of more than 20 new, hitherto unknown acetylation, methylation, or phosphorylation sites in the mammalian core histones by only one straightforward experiment [375] (Fig. 21). For this purpose the core histones isolated from calf thymus nuclei were HPLC purified and aliquots of each fraction were digested with the proteases pepsin, trypsin, V8DE, and V8E. Each aliquot was analyzed with ESI-FT-ICR MS, the measured peptide masses were compared with the expected masses, and modified species were assigned based on mass differences that match acetylation of a lysine residue, single, double, or triple methylation of a lysine or arginine residue, or phosphorylation of a serine or threonine residue. The results were confirmed and the location of the modifications assigned to individual amino acid residues by comparing the sequence of the assigned peptides across the five different digests of each fraction. In this paper it was also discovered that methylation of lysine 59 of the central region of histone H4 is essential for transcriptional silencing at the yeast silent loci and telomeres.

Within the last 5 years, by the use of MS the number of known PTMs of the histone protein family has grown dramatically and not two months pass without another previously unknown modification being published. In addition to PMF (Fig. 7), MS/MS-based techniques (Fig. 8) including topdown analyses of intact histone molecular ions [59, 61, 377] (Fig. 22) have been extensively and successfully applied for the identification of new histone PTMs as well as quantification of their abundance [60, 378, 379]. More

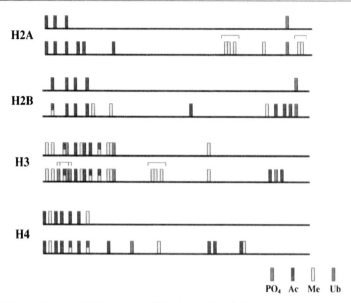

Fig. 21 Schematic map of histone modifications identified in a single study by peptide mass fingerprinting [375]. Each core histone is represented by a pair of *horizontal lines*. The *color-coded bars* on the *top line* indicate the sites of previously identified modifications, while the new sites of modification identified in this study have been added to the *bottom line*. Sites enclosed within brackets indicate modifications that could not be definitively localized to a specific residue. Reproduced from [375]

detailed information on this development and how the different PTMs were found and identified by MS is provided elsewhere [60, 374, 376, 380]. Regularly updated maps that show known methylation, acetylation, phosphorylation, and ubiquitinylation sites for the individual histones are provided at: http://www.histone.com/modification_map.htm.

It has also been realized that histone modification patterns and levels are an important aspect when studying diseased gene expression states. For instance, it has recently been shown that certain aberrant posttranslational histone modifications are associated with the pathogenesis of the autoimmune disease systemic lupus erythematosus (SLE) [381]. In this study, MS combined with stable isotope labeling was used to differentially analyze histone modifications in splenocytes from a mouse model of SLE. Compared to the control, the disease model showed a global site-specific hypermethylation, except for H3K4, and hypoacetylation in histone H3 and H4.

In vivo administration of the histone deacetylase inhibitor trichostatin A corrected the site-specific hypoacetylation states on H3 and H4 and improved the phenotype of the disease. In this study, novel histone modifications such as H3K18 methylation, H4K31 methylation, and H4K31 acetylation were also discovered and found to be differentially expressed in the disease

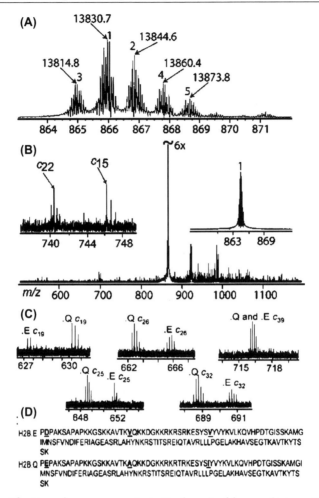

Fig. 22 Identification of sequence variants (isoforms) of human histone H2B isolated from asynchronously grown HeLa S3 cells by ESI-FT-ICR MS and MS/MS. To minimize sample heterogeneity (partial oxidation during the sample preparation), both methionine residues of H2B were quantitatively oxidized with performic acid prior to MS. **a** Broadband mass spectrum of the H2B isolate after RP HPLC purification revealed five main isotopic distributions, which were afterward analyzed by top-down MS. The determined oxidized monoisotopic mass is listed above each molecular ion species carrying 16 positive charges. Using ECD for direct MS/MS analysis, the H2B isoforms H2B.K and H2B.T (both 13 814.5 Da), H2B.J (13 816.5 Da), H2B.A (13 830.5 Da), H2B.Q and H2B.E (both 13 844.5 Da), H2B.B (13 860.5 Da), H2B.F (13874.5), and monoacetylated H2B.A (13 872.5 Da) were identified. The masses include oxidation of both methionine residues to their sulfones (+64 Da). **b** Identification of the isobaric H2B sequence variants H2B.Q and H2B.E by ECD MS/MS of peak 2 (13 844.6 Da). **c** Key fragment ions in the 625–725 m/z region reporting on the presence of H2B.E and H2B.Q. **d** Sequence alignment of H2B.E and H2B.Q. Sequence differences are underlined. Reproduced from [59] with permission from © American Chemical Society, 2006

model. Another noteworthy observation concerns aging of mammals. Using hydrophilic-interaction LC-MS, it was found that the level of trimethylation of H4K20 is significantly increased in the kidney and liver of rats older than 30 days, whereas the dominant dimethylated form did not essentially change from young to old animals [382].

The question of whether a histone code consisting of combinations of different histone PTMs that correlate with specific chromatin states really exists or not has been and still is heavily debated among scientists, and many experiments, including thousands of ESI and MALDI MS measurements, were performed to find evidence that the different PTMs of the histones and their biological functions are logically interconnected. For instance, by combining metal-affinity chromatography, MS/MS, and immunoassay methods to characterize histone H3 purified from mitotically arrested HeLa cells, it was shown that phosphorylation of Ser10, Ser28, and Thr3 of histone H3 is reduced when adjacent Lys residues are methylated [383]. This observation supports the existence of a "methyl/phos" binary switch [384]. Today, it is clear that certain patterns of histone PTMs correlate with distinct chromosomal states that regulate access to DNA. What is not clear is whether these PTMs are mostly introduced on the chromatin template, which could be regarded as "code writing", or instead on soluble histones before assembly, which could be interpreted as a structural prerequisite [385]. Implementation of a code, as commonly understood, also requires a means to read (translate) the encoded information. The assumption that this part is executed by a set of proteins that recognize and bind to specific histone modifications and thereby, in concert, induce formation of a specific chromatin state, is equally uncertain. Nevertheless, whether a histone code exists or not, there is no doubt that PTMs play an important role in the regulation of chromatin dynamics and gene expression [386].

Chromatin states can be distinguished by differential PTMs or by utilization of different histone isoforms and sequence variants (subtypes) [387]. These are encoded by different genes, e.g., the human genome contains 13 H2A genes encoding at least six different amino acid sequences and 15 H2B genes encoding 11 different H2B subtypes. These differ remarkably little, e.g., by the identity of only one or two amino acid residues [388]. Therefore, to assign modification sites to the correct histone variant and quantify their presence in a given sample is a true challenge for MS [379, 389–391]. In one study the relative abundance of H3 and H3.3 and their lysine modifications was quantified [389]. Using a *Drosophila* cell line system, in which H3.3 has been shown to specifically package active loci, it could be shown that H3.3 accounts for about 25% of total histone 3 in bulk chromatin, sufficient to package essentially all actively transcribed genes. MS and antibody characterization of separated histone 3 fractions revealed that H3.3 is relatively enriched in modifications associated with transcriptional activity and deficient in dimethyl lysine-9, which is abundant in heterochromatin.

In a global analysis of histone H2A and H2B variants derived from Jurkat cells by MS, nine histone H2A and 11 histone H2B subtypes were identified, some of which had only been postulated before at the DNA level [391]. This was achieved by combining MS with HPLC separations and enzymatic proteolysis using endoproteinase Glu-C, endoproteinase Arg-C, and trypsin. With regard to modification status, e.g., the two main H2A variants, H2A.o and H2A.c, as well as H2A.l were found either acetylated at Lys-5 or phosphorylated at Ser-1. For the replacement histone H2A.z, acetylation at Lys-4 and Lys-7 was observed. The main histone H2B variant, H2B.a, was found acetylated at Lys-12, -15, and -20. In an other study, a direct, top-down MS approach was applied to identify H2B isoforms isolated from asynchronous HeLa cells using ESI-FT-ICR MS and ECD for MS/MS analysis of intact protein molecular ions (Fig. 22) [59]. These cells were found to express H2B.A, H2B.B, H2B.E, H2B.F, H2B.J, H2B.K, H2B.Q, and H2B.T.

5.3
Higher-Order Structures and Interactions

MS has been used to characterize the structures of DNA-binding proteins in many different ways. An early example that demonstrates the power of ESI MS for the study of noncovalent protein–DNA complexes is the determination of the stoichiometry of the interactions of protein V and the PU.1 transcription factor with single- and double-stranded DNA [263, 264]. 3D structural information for the transcription factor Max was generated by MS-based protein footprinting [165] (Fig. 23). In the absence of DNA, Max was rapidly degraded, suggesting a flexible structure. In contrast, in the presence of a double-stranded DNA sequence to which Max binds specifically, the proteolysis rate was reduced by two orders of magnitude. Specifically, as expected, the N-terminus was observed to be highly protected by tight interactions with the DNA probe. Another example is the structure and topology of the thyroid transcription factor 1 homeodomain bound to DNA, which was studied by ESI and MALDI MS in combination with limited proteolysis as well as surface labeling using acetylation of lysine residues as the modification reaction [166]. The results of these experiments were consistent with a 3D structure previously established by NMR. A third example of the application of MS-based protein footprinting is the determination of structural linker sequences of the NtrC transcription factor of *E. coli* [167]. Limited proteolysis combined with MALDI MS has also been used to study the effect of known cancer-associated mutations in the DNA-binding protein BRCA1 [170]. For this purpose, the heterodimer BRCA1-BARD1 was analyzed and the experimental results showed that the cancer-related mutations alter the protein–protein interaction by changing the structure of a binding loop in BRAC1 that interacts with BARD1.

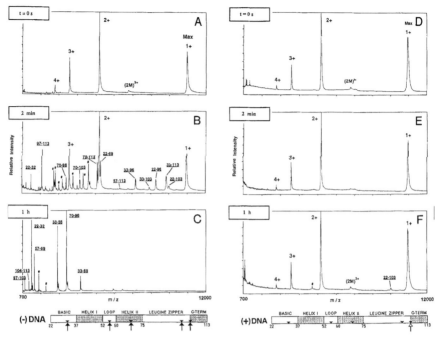

Fig. 23 Characterization of the transcription factor Max by MS-based protein footprinting. MALDI mass spectra of the products of a time-course V8 proteolysis of Max in the **a–c** absence and **d–f** presence of Max-specific DNA. Digests were performed in 15 mM KCl buffer (pH 6) at 25 °C. Three time points are shown: **a**, **d** 0 s (before the addition of V8); **b**, **e** 2 min of digestion; and **c**, **f** 1 h of digestion. Peaks labeled Max, 1+ arise from singly protonated Max and those labeled 2+, 3+, and 4+ arise from multiply protonated Max, resulting from the MALDI process. The peak labeled $(2M)^{3+}$ arises from triply protonated Max dimers, formed during the MALDI process. Peaks corresponding to singly charged V8 fragments are labeled with their sequence as determined from their measured mass. For clarity only the singly charged fragment peaks are labeled with sequences; the corresponding multiply charged fragment peaks are labeled with #. Linear diagrams at the *bottom* of the figure summarize the progress of proteolysis following 1 h of V8 digestion in the absence (**c**) and presence (**f**) of Max-specific DNA. *Small solid arrowheads* inside the diagrams point to the sites of rapid cleavage by V8 that are observed in the absence of DNA. Rapid cleavage occurs at the five glutamate residues 32, 56, 69, 96, and 103. *Large arrows* outside the diagram point to the observed sites of proteolysis determined by MALDI MS: *dark shaded arrows* signify a rapid and complete cleavage (minutes to hours); *open arrows* signify a slow cleavage (hours to days). Reproduced from [165] with permission from © The Protein Society, 2006

A different strategy was used to characterize the structure of the ferric-uptake regulation factor Fur [174]. As a first approach, the reactivity (accessibility) of each lysine residue was investigated by chemical modification followed by ESI MS. After metal activation, K76 was found to be protected from the modification reaction in the presence of DNA. The conformational changes induced upon metal binding were then characterized by HX experiments. Based on these results and the predicted secondary structure, Fur was classified as a nonclassical helix-turn-helix protein with structural homology to the diphtheria toxin repressor. To study the conformational dynamics of the heat shock transcription factor σ32, HX MS proved particularly valuable because NMR was not an option due to the strong tendency of σ32 to aggregate at high concentrations [204]. To clarify whether σ32 acts as a thermosensor, its folded states were studied at 37 and 42 °C, providing optimal growth or heat-stress conditions, respectively. The results suggested a high degree of protein flexibility at normal temperature and a reversible unfolding of a small structural motif under heat stress. The location of this motif was identified by pepsin proteolysis followed by MS. There have been many examples in recent years of HX MS being combined with high-resolution structures to explore the organization and dynamics of complex molecular assemblies. A good example is the study of the effect of DNaseI ligand interactions with G-actin on the polymerization of G-actin to form F-actin [392].

An impressive example of the power of combining immunoaffinity copurification with MS for the identification of the interactions of DNA-binding proteins with other proteins is the yeast TATA box binding transcription factor TFIID machinery [393]. To gain insight into TfIID function, a proteomic catalog of proteins specifically interacting with TFIID subunits was established using polyclonal antibodies directed against each subunit. The copurified proteins were identified by 2D-LC-MS/MS and based on these data a number of novel protein–protein associations could be assigned. Several of these were subsequently characterized in detail, including interactions between TATA box binding proteins and the RSC chromatin remodeling complex, and the TAF17p-dependent association of the Swi6p transactivator protein with TfIID. In addition, three novel subunits of the SAGA acetyl transferase complex were identified, including a putative ubiquitin-specific protease component.

To demonstrate the potential of combining SPR-based BIA with MALDI-TOF MS for the characterization of DNA–protein interactions, the binding of *E. coli*'s transcription factor ParR to its parA promoter partition site parC of plasmid R1 was monitored and quantified by BIA in the presence of a molar excess of insulin (porcine), myoglobin (horse), and beta lactoglobulin (variants A and B, bovine milk) serving as unrelated protein background. After washing, affinity-bound proteins were eluted and analyzed by MALDI-TOF MS [286]. This measurement confirmed that the affinity purification was successful and that the protein amount quantified by SPR was not significantly contaminated by any of the background proteins.

An important feature of genetic tagging for affinity purification followed by MS is that it can easily be scaled up to large numbers of proteins, thus enabling systematic exploration of protein–protein interaction networks. This was impressively demonstrated for 589 TAP-tagged proteins in yeast. Based on the identified binding partners, an interaction map was drawn that encompasses 232 meaningful complexes and 1440 distinct gene products including many DNA-binding proteins [394]. An interesting result of this study was that, in contrast to other large-scale techniques for the systematic identification of protein–protein interactions, the error rate was below 20%. In comparison, for a previous large-scale flag tag (single tag) based screen a 50% error rate was expected [270], and for two global yeast two-hybrid screens the error was rated at 45–80% [395–397]. This enormous improvement was possible because only in the TAP experiments were the tagged proteins expressed

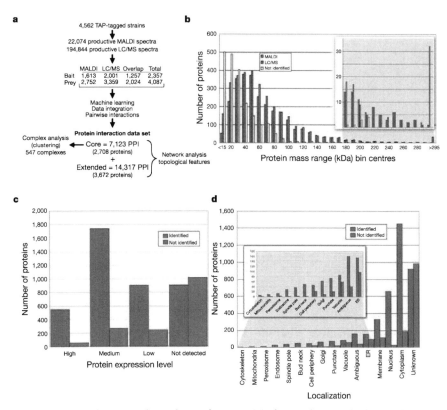

Fig. 24 MS-based large-scale analysis of *S. cerevisiae*'s protein–protein interaction network (interactome). Tandem affinity purification (TAP) was used to isolate the interaction partners of 4562 different tagged proteins. Each sample was analyzed twice, by SDS-PAGE-MALDI MS and by LC-ESI MS/MS, to increase coverage and accuracy. In total 7123 protein–protein interactions were identified involving 2708 different proteins and 547 protein complexes. **a** Summary of the experimental strategy and data analysis. PPI, protein–protein interactions. **b–f** The proportions of proteins identified as bait or prey are shown in relation to protein mass (**b**), and known expression level (**c**) and intracellular localization (**d**). Reproduced from [399] with permission from © Nature Publishing Group, 2006

under their endogenous promoters. For comparison, in the flag-tag experiments all tagged proteins were overexpressed with the inevitable high risk of false positive results. Notably, the yeast protein–protein interaction network discussed above has recently been further extended significantly [398, 399]. In one large-scale experiment, 7123 protein–protein interactions were identified involving 2708 different proteins and 547 protein complexes (Fig. 24). To achieve this improvement, instead of 589, 4562 different tagged proteins of yeast *S. cerevisiae* were used for the TAP-MS-based identification of protein–protein interactions [399].

Acknowledgements The authors thank Dr. J. Gobom, Max Planck Institute for Molecular Genetics, for assistance in preparing the manuscript and scientific discussion. Funding of the Zukunftsfond of Berlin within the project 2D/3D-ProteinChips is acknowledged.

References

1. Fenn JB, Mann M, Meng CK, Wong SF, Whitehouse CM (1989) Science 246:64
2. Fenn JB, Mann M, Meng CK, Wong SF, Whitehouse CM (1990) Mass Spectrom Rev 9:37
3. Karas M, Hillenkamp F (1988) Anal Chem 60:2299
4. Karas M, Bahr U, Hillenkamp F (1989) Int J Mass Spectrom Ion Process 92:231
5. Karas M, Bahr U, Ingendoh A, Hillenkamp F (1989) Angew Chem Int Ed Engl 28:760
6. Pandey A, Mann M (2000) Nature 405:837
7. Aebersold R, Mann M (2003) Nature 422:198
8. Tyers M, Mann M (2003) Nature 422:193
9. Patterson SD, Aebersold RH (2003) Nat Genet 33:311
10. Lane CS (2005) Cell Mol Life Sci 62:848
11. Domon B, Aebersold R (2006) Science 312:212
12. Zenobi R, Knochenmuss R (1998) Mass Spectrom Rev 17:337
13. Aebersold R, Goodlett DR (2001) Chem Rev 101:269
14. Krutchinsky AN, Kalkum M, Chait BT (2001) Anal Chem 73:5066
15. Loboda AV, Krutchinsky AN, Bromirski M, Ens W, Standing KG (2000) Rapid Commun Mass Spectrom 14:1047
16. Elias JE, Haas W, Faherty BK, Gygi SP (2005) Nat Methods 2:667
17. Baldwin MA, Medzihradszky KF, Lock CM, Fisher B, Settineri TA, Burlingame AL (2001) Anal Chem 73:1707
18. Steen H, Kuster B, Mann M (2001) J Mass Spectrom 36:782
19. Amster IJ (1996) J Mass Spectrom 31:1325
20. Marshall AG, Hendrickson CL, Jackson GS (1998) Mass Spectrom Rev 17:1
21. Zabrouskov V, Senko MW, Du Y, Leduc RD, Kelleher NL (2005) J Am Soc Mass Spectrom 16:2027
22. Syka JEP, Marto JA, Bai DL, Horning S, Senko MW, Schwartz JC, Ueberheide B, Garcia B, Busby S, Muratore T, Shabanowitz J, Hunt DF (2004) J Proteome Res 3:621
23. Olsen JV, Mann M (2004) Proc Natl Acad Sci USA 101:13417
24. Olsen JV, de Godoy LMF, Li GQ, Macek B, Mortensen P, Pesch R, Makarov A, Lange O, Horning S, Mann M (2005) Mol Cell Proteomics 4:2010
25. Hager JW (2002) Rapid Commun Mass Spectrom 16:512
26. Schwartz JC, Senko MW, Syka JEP (2002) J Am Soc Mass Spectrom 13:659

27. Yates JR, Cociorva D, Liao LJ, Zabrouskov V (2006) Anal Chem 78:493
28. McLafferty FW (1997) J Am Soc Mass Spectrom 8:1
29. McLafferty FW, Kelleher NL, Begley TP, Fridriksson EK, Zubarev RA, Horn DM (1998) Curr Opin Chem Biol 2:571
30. Thalassinos K, Slade SE, Jennings KR, Scrivens JH, Giles K, Wildgoose J, Hoyes J, Bateman RH, Bowers MT (2004) Int J Mass Spectrom 236:55
31. Ruotolo BT, Gillig KJ, Woods AS, Egan TF, Ugarov MV, Schultz JA, Russell DH (2004) Anal Chem 76:6727
32. McLean JA, Ruotolo BT, Gillig KJ, Russell DH (2005) Int J Mass Spectrom 240:301
33. Myung S, Wiseman JM, Valentine SJ, Takats Z, Cooks RG, Clemmer DE (2006) J Phys Chem B 110:5045
34. Baumketner A, Bernstein SL, Wyttenbach T, Bitan G, Teplow DB, Bowers MT, Shea JE (2006) Protein Sci 15:420
35. Ruotolo BT, McLean JA, Gillig KJ, Russell DH (2004) J Mass Spectrom 39:361
36. Kussmann M, Nordhoff E, Nielsen HR, Haebel S, RosselLarsen M, Jakobsen L, Gobom J, Mirgorodskaya E, lKristensen AK, Palm L, Roepstorff P (1997) J Mass Spectrom 32:593
37. Ong SE, Mann M (2005) Nat Chem Biol 1:252
38. Stapels MD, Barofsky DF (2004) Anal Chem 76:5423
39. Spengler B (1997) J Mass Spectrom 32:1019
40. Vestal ML, Campbell JM (2005) Biol Mass Spectrom 402:79
41. Medzihradszky KF, Campbell JM, Baldwin MA, Falick AM, Juhasz P, Vestal ML, Burlingame AL (2000) Anal Chem 72:552
42. Suckau D, Resemann A, Schuerenberg M, Hufnagel P, Franzen J, Holle A (2003) Anal Bioanal Chem 376:952
43. Zhang WZ, Krutchinsky AN, Chait BT (2003) J Am Soc Mass Spectrom 14:1012
44. Little DP, Speir JP, Senko MW, Oconnor PB, Mclafferty FW (1994) Anal Chem 66:2809
45. Zhai H, Han XM, Breuker K, McLafferty FW (2005) Anal Chem 77:5777
46. Cooper HJ, Hakansson K, Marshall AG (2005) Mass Spectrom Rev 24:201
47. McFarland MA, Chalmers MJ, Quinn JP, Hendrickson CL, Marshall AG (2005) J Am Soc Mass Spectrom 16:1060
48. Medzihradszky KF (2005) Biol Mass Spectrom 402:209
49. Syka JEP, Coon JJ, Schroeder MJ, Shabanowitz J, Hunt DF (2004) Proc Natl Acad Sci USA 101:9528
50. Hartmer R, Lubeck M (2005) LC GC Europe, p 11
51. Coon JJ, Shabanowitz J, Hunt DF, Syka JEP (2005) J Am Soc Mass Spectrom 16:880
52. Pitteri SJ, Chrisman PA, McLuckey SA (2005) Anal Chem 77:5662
53. Zubarev RA, Kelleher NL, McLafferty FW (1998) J Am Chem Soc 120:3265
54. Zubarev RA, Kruger NA, Fridriksson EK, Lewis MA, Horn DM, Carpenter BK, McLafferty FW (1999) J Am Chem Soc 121:2857
55. Zubarev RA, Horn DM, Fridriksson EK, Kelleher NL, Kruger NA, Lewis MA, Carpenter BK, McLafferty FW (2000) Anal Chem 72:563
56. Kruger NA, Zubarev RA, Horn DM, McLafferty FW (1999) Int J Mass Spectrom 187:787
57. Kruger NA, Zubarev RA, Carpenter BK, Kelleher NL, Horn DM, McLafferty FW (1999) Int J Mass Spectrom 183:1
58. Horn DM, Ge Y, McLafferty FW (2000) Anal Chem 72:4778
59. Siuti N, Roth MJ, Mizzen CA, Kelleher NL, Pesavento JJ (2006) J Proteome Res 5:233
60. Thomas CE, Kelleher NL, Mizzen CA (2006) J Proteome Res 5:240
61. Boyne MT, Pesavento JJ, Mizzen CA, Kelleher NL (2006) J Proteome Res 5:248

62. Coon JJ, Ueberheide B, Syka JEP, Dryhurst DD, Ausio J, Shabanowitz J, Hunt DF (2005) Proc Natl Acad Sci USA 102:9463
63. Karas M, Ingendoh A, Bahr U, Hillenkamp F (1989) Biomed Environ Mass Spectrom 18:841
64. Strupat K, Karas M, Hillenkamp F (1991) Int J Mass Spectrom Ion Process 111:89
65. Gobom J, Nordhoff E, Mirgorodskaya E, Ekman R, Roepstorff P (1999) J Mass Spectrom 34:105
66. Ishihama Y, Rappsilber J, Andersen JS, Mann M (2002) J Chrom A 979:233
67. Rappsilber J, Ishihama Y, Mann M (2003) Anal Chem 75:663
68. Ishihama Y, Rappsilber J, Mann M (2006) J Proteome Res 5:988
69. Wilm MS, Mann M (1994) Int J Mass Spectrom Ion Process 136:167
70. Wilm M, Shevchenko A, Houthaeve T, Breit S, Schweigerer L, Fotsis T, Mann M (1996) Nature 379:466
71. Shevchenko A, Chernushevich I, Ens W, Standing KG, Thomson B, Wilm M, Mann M (1997) Rapid Commun Mass Spectrom 11:1015
72. Zhang S, Van Pelt CK (2004) Expert Rev Proteomics 1:449
73. Washburn MP, Wolters D, Yates JR (2001) Nat Biotechnol 19:242
74. Washburn MP, Ulaszek RR, Yates JR (2003) Anal Chem 75:5054
75. Vorm O, Roepstorff P, Mann M (1994) Anal Chem 66:3281
76. Schuerenbeg M, Luebbert C, Eickhoff H, Kalkum M, Lehrach H, Nordhoff E (2000) Anal Chem 72:3436
77. Gobom J, Schuerenberg M, Mueller M, Theiss D, Lehrach H, Nordhoff E (2001) Anal Chem 73:434
78. Nordhoff E, Schurenberg M, Thiele G, Lubbert C, Kloeppel KD, Theiss D, Lehrach H, Gobom J (2003) Int J Mass Spectrom 226:163
79. Hutchens TW, Yip TT (1993) Rapid Commun Mass Spectrom 7:576
80. Merchant M, Weinberger SR (2000) Electrophoresis 21:1164
81. Seibert V, Wiesner A, Buschmann T, Meuer J (2004) Pathol Res Pract 200:83
82. Reddy G, Dalmasso EA (2003) J Biomed Biotechnol 2003:237
83. Tang N, Tornatore P, Weinberger SR (2004) Mass Spectrom Rev 23:34
84. Clarke CH, Buckley JA, Fung ET (2005) Clin Chem Lab Med 43:1314
85. Bane TK, LeBlanc JF, Lee TD, Riggs AD (2002) Nucleic Acids Res 30:e69
86. Forde CE, Gonzales AD, Smessaert JM, Murphy GA, Shields SJ, Fitch JP, McCutchen-Maloney SL (2002) Biochem Biophys Res Commun 290:1328
87. Nordhoff E, Kirpekar F, Roepstorff P (1996) Mass Spectrom Rev 15:67
88. Limbach PA (1996) Mass Spectrom Rev 15:297
89. Jurinke C, Oeth P, van den Boom D (2004) Mol Biotechnol 26:147
90. Henzel WJ, Billeci TM, Stults JT, Wong SC, Grimley C, Watanabe C (1993) Proc Natl Acad Sci USA 90:5011
91. James P, Quadroni M, Carafoli E, Gonnet G (1993) Biochem Biophys Res Commun 195:58
92. Mann M, Hojrup P, Roepstorff P (1993) Biol Mass Spectrom 22:338
93. Pappin DJC, Hojrup P, Bleasby AJ (1993) Curr Biol 3:327
94. Yates JR, Speicher S, Griffin PR, Hunkapiller T (1993) Anal Biochem 214:397
95. Egelhofer V, Bussow K, Luebbert C, Lehrach H, Nordhoff E (2000) Anal Chem 72:2741
96. Eng JK, Mccormack AL, Yates JR (1994) J Am Soc Mass Spectrom 5:976
97. Griffin PR, MacCoss MJ, Eng JK, Blevins RA, Aaronson JS, Yates JR (1995) Rapid Commun Mass Spectrom 9:1546
98. Yates JR, Eng JK, Clauser KR, Burlingame AL (1996) J Am Soc Mass Spectrom 7:1089

99. Yates JR, Eng JK, Mccormack AL (1995) Anal Chem 67:3202
100. Yates JR, Eng JK, Mccormack AL, Schieltz D (1995) Anal Chem 67:1426
101. Yates JR, Link AJ, Schieltz D, Hays L, Carmack E, Eng J (1997) Abstr Pap Am Chem Soc 214:110
102. Yates JR, McCormack AL, Eng J (1996) Anal Chem 68:A534
103. Perkins DN, Pappin DJC, Creasy DM, Cottrell JS (1999) Electrophoresis 20:3551
104. Steen H, Mann M (2004) Nat Rev Mol Cell Biol 5:699
105. Spengler B (2004) J Am Soc Mass Spectrom 15:703
106. Carte N, Cavusoglu N, Leize E, Van Dorsselaer A, Charlet M, Bulet P (2001) Eur J Mass Spectrom 7:399
107. Dancik V, Addona TA, Clauser KR, Vath JE, Pevzner PA (1999) J Comput Biol 6:327
108. Zhang ZQ, McElvain JS (2000) Anal Chem 72:2337
109. Shevchenko A, Wilm M, Mann M (1997) J Prot Chem 16:481
110. Shevchenko A, Chernushevich I, Shevchenko A, Wilm M, Mann M (2002) Mol Biotechnol 20:107
111. Mann M, Wilm M (1994) Anal Chem 66:4390
112. Gruhler A, Olsen JV, Mohammed S, Mortensen P, Faergeman NJ, Mann M, Jensen ON (2005) Mol Cell Proteomics 4:310
113. Spengler B, Luetzenkirchen F, Metzger S, Chaurand P, Kaufmann R, Jeffery W, Bartlet-Jones M, Pappin DJC (1997) Int J Mass Spectrom 169:127
114. Reid GE, Roberts KD, Simpson RJ, O'Hair RAJ (2005) J Am Soc Mass Spectrom 16:1131
115. Pashkova A, Chen HS, Rejtar T, Zang X, Giese R, Andreev V, Moskovets E, Karger BL (2005) Anal Chem 77:2085
116. Cagney G, Emili A (2002) Nat Biotechnol 20:163
117. Strittmatter EF, Kangas LJ, Petritis K, Mottaz HM, Anderson GA, Shen YF, Jacobs JM, Camp DG, Smith RD (2004) J Proteome Res 3:760
118. Pasa-Tolic L, Masselon C, Barry RC, Shen YF, Smith RD (2004) Biotechniques 37:621
119. Zimmer JSD, Monroe ME, Qian WJ, Smith RD (2006) Mass Spectrom Rev 25:450
120. Patrie SM, Ferguson JT, Robinson DE, Whipple D, Rother M, Metcalf WW, Kelleher NL (2006) Mol Cell Proteomics 5:14
121. Sze SK, Ge Y, Oh H, McLafferty FW (2002) Proc Natl Acad Sci USA 99:1774
122. Ge Y, Lawhorn BG, El Naggar M, Strauss E, Park JH, Begley TP, McLafferty FW (2002) J Am Chem Soc 124:672
123. Kelleher NL, Lin HY, Valaskovic GA, Aaserud DJ, Fridriksson EK, McLafferty FW (1999) J Am Chem Soc 121:806
124. Hunt DF, Yates JR, Shabanowitz J, Winston S, Hauer CR (1986) Proc Natl Acad Sci USA 83:6233
125. Horn DM, Zubarev RA, McLafferty FW (2000) Proc Natl Acad Sci USA 97:10313
126. Johnson RS, Walsh KA (1992) Protein Sci 1:1083
127. Haebel S, Jensen C, Andersen SO, Roepstorff P (1995) Protein Sci 4:394
128. Buzy A, Gagnon J, Lamy J, Thibault P, Forest E, Hudryclergeon G (1995) Eur J Biochem 233:93
129. Bauer MD, Sun YP, Wang F (1999) J Prot Chem 18:337
130. Gevaert K, Vandekerckhove J (2000) Electrophoresis 21:1145
131. Ostrom PH, Gandhi H, Strahler JR, Walker AK, Andrews PC, Leykam J, Stafford TW, Kelly RL, Walker DN, Buckley M, Humpula J (2006) Geochim Cosmochim Acta 70:2034
132. Standing KG (2003) Curr Opin Struct Biol 13:595

133. Baldwin MA (2004) Mol Cell Proteomics 3:1
134. Jungblut P, Thiede B (1997) Mass Spectrom Rev 16:145
135. Kuster B, Mortensen P, Andersen JS, Mann M (2001) Proteomics 1:641
136. Shevchenko A, Jensen ON, Podtelejnikov AV, Sagliocco F, Wilm M, Vorm O, Mortensen P, Shevchenko A, Boucherie H, Mann M (1996) Proc Natl Acad Sci USA 93:14440
137. Mann M, Pandey A (2001) Trends Biochem Sci 26:54
138. Turner BM (1993) Cell 75:5
139. Strahl BD, Allis CD (2000) Nature 403:41
140. Smith CM, Haimberger ZW, Johnson CO, Wolf AJ, Gafken PR, Zhang ZL, Parthun MR, Gottschling DE (2002) Proc Natl Acad Sci USA 99:16454
141. Turner BM (2005) Nat Struct Mol Biol 12:110
142. Hamdan M, Galvani M, Righetti PG (2001) Mass Spectrom Rev 20:121
143. Mann M, Jensen ON (2003) Nat Biotechnol 21:255
144. Mann M, Ong SE, Gronborg M, Steen H, Jensen ON, Pandey A (2002) Trends Biotechnol 20:261
145. Hendrickson CL, Emmett MR, Nilsson CL, Marshall AG (2004) Abstr Pap Am Chem Soc 228:U157
146. Salih E (2005) Mass Spectrom Rev 24:828
147. Medzihradszky KF (2005) Methods Enzymol 405:116
148. Stults JT, Arnott D (2005) Biol Mass Spectrom 402:245
149. Blagoev B, Ong SE, Kratchmarova I, Mann M (2004) Nat Biotechnol 22:1139
150. Kokubu M, Ishihama Y, Sato T, Nagasu T, Oda Y (2005) Anal Chem 77:5144
151. Carr SA, Annan RS, Huddleston MJ (2005) Methods Enzymol 405:82
152. Steen H, Kuster B, Fernandez M, Pandey A, Mann M (2001) Anal Chem 73:1440
153. Kocher T, Savitski MM, Nielsen ML, Zubarev RA (2006) J Proteome Res 5:659
154. Hanson CL, Robinson CV (2004) J Biol Chem 279:24907
155. Videler H, Ilag LL, McKay ARC, Hanson CL, Robinson CV (2005) FEBS Lett 579:943
156. Heck AJR, van den Heuvel RHH (2004) Mass Spectrom Rev 23:368
157. van den Heuvel RH, Heck AJR (2004) Curr Opin Chem Biol 8:519
158. Aplin RT, Robinson CV, Schofield CJ, Westwood NJ (1994) J Chem Soc Chem Comm 2415
159. Beck JL, Colgrave ML, Ralph SF, Sheil MM (2001) Mass Spectrom Rev 20:61
160. Hofstadler SA, Griffey RH (2001) Chem Rev 101:377
161. Craig TA, Benson LM, Tomlinson AJ, Veenstra TD, Naylor S, Kumar R (1999) Nat Biotechnol 17:1214
162. Loo JA (1997) Mass Spectrom Rev 16:1
163. Xu NX, Pasa-Tolic L, Smith RD, Ni SS, Thrall BD (1999) Anal Biochem 272:26
164. Forde CE, McCutchen-Maloney SL (2002) Mass Spectrom Rev 21:419
165. Cohen SL, Ferredamare AR, Burley SK, Chait BT (1995) Protein Sci 4:1088
166. Scaloni A, Monti M, Acquaviva R, Tell G, Damante G, Formisano S, Pucci P (1999) Biochemistry 38:64
167. Bantscheff M, Weiss V, Glocker MO (1999) Biochemistry 38:11012
168. Veenstra TD, Johnson KL, Tomlinson AJ, Craig TA, Kumar R, Naylor S (1998) J Am Soc Mass Spectrom 9:8
169. Bennett KL, Kussmann M, Bjork P, Godzwon M, Mikkelsen M, Sorensen P, Roepstorff P (2000) Protein Sci 9:1503
170. Brzovic PS, Meza JE, King MC, Klevit RE (2001) J Biol Chem 276:41399
171. Imre T, Zsila F, Szabo PT (2003) Rapid Commun Mass Spectrom 17:2464
172. Jawhari A, Boussert S, Lamour V, Atkinson RA, Kieffer B, Poch O, Potier N, van Dorsselaer A, Moras D, Poterszman A (2004) Biochemistry 43:14420

173. Nordhoff E, Roepstorff P (1998) Proceedings of the 46th ASMS conference on mass spectrometry and allied topics, Orlando, Florida, 31 May–4 June 1998. American Society for Mass Spectrometry, Orlando, p 388
174. de Peredo AG, Saint-Pierre C, Latour JM, Michaud-Soret I, Forest E (2001) J Mol Biol 310:83
175. Alexander P, Moroson H (1962) Nature 194:882
176. Smith KC (1962) Biochem Biophys Res Commun 8:157
177. Shaw AA, Falick AM, Shetlar MD (1992) Biochemistry 31:10976
178. Shivanna BD, Mejillano MR, Williams TD, Himes RH (1993) J Biol Chem 268:127
179. Steen H, Jensen ON (2002) Mass Spectrom Rev 21:163
180. Orlando V, Strutt H, Paro R (1997) Methods 11:205
181. Kaufman BA, Newman SM, Hallberg RL, Slaughter CA, Perlman PS, Butow RA (2000) Proc Natl Acad Sci USA 97:7772
182. Urlaub H, Kruft V, Bischof O, Muller EC, Wittmannliebold B (1995) EMBO J 14:4578
183. Urlaub H, Thiede B, Muller EC, WittmannLiebold B (1997) J Prot Chem 16:375
184. Sastry SS (1996) Biochemistry 35:13519
185. Sastry S, Ross BM (1998) Proc Natl Acad Sci USA 95:9111
186. Meisenheimer KM, Koch TH (1997) Crit Rev Biochem Mol Biol 32:101
187. Norris CL, Meisenheimer PL, Koch TH (1996) J Am Chem Soc 118:5796
188. Jensen ON, Barofsky DF, Young MC, Vonhippel PH, Swenson S, Seifried SE (1993) Rapid Commun Mass Spectrom 7:496
189. Nordhoff E, Cramer R, Karas M, Hillenkamp F, Kirpekar F, Kristiansen K, Roepstorff P (1993) Nucleic Acids Res 21:3347
190. Nordhoff E, Kirpekar F, Karas M, Cramer R, Hahner S, Hillenkamp F, Kristiansen K, Roepstorff P, Lezius A (1994) Nucleic Acids Res 22:2460
191. Marie G, Serani L, Laprevote O, Cahuzac B, Guittet E, Felenbok B (2001) Protein Sci 10:99
192. Akashi S, Niitsu U, Yuji R, Ide H, Hirayama K (1993) Biol Mass Spectrom 22:124
193. Taylor IA, Webb M, Kneale GG (1996) J Mol Biol 258:62
194. Fiedler W, Borchers C, Macht M, Deininger SO, Przybylski M (1998) Bioconjug Chem 9:236
195. Hoofnagle AN, Resing KA, Ahn NG (2003) Annu Rev Biophys Biomol Struct 32:1
196. Eyles SJ, Kaltashov IA (2004) Methods 34:88
197. Kaltashov IA, Eyles SJ (2002) J Mass Spectrom 37:557
198. Krishna MMG, Hoang L, Lin Y, Englander SW (2004) Methods 34:51
199. Wales TE, Engen JR (2006) Mass Spectrom Rev 25:158
200. Garcia RA, Pantazatos D, Villarreal FJ (2004) Assay Drug Dev Technol 2:81
201. Chung EW, Nettleton EJ, Morgan CJ, Gross M, Miranker A, Radford SE, Dobson CM, Robinson CV (1997) Protein Sci 6:1316
202. Wang AJ, Englander SW (1996) Curr Opin Biotechnol 7:403
203. Hildebrandt P, Vanhecke F, Buse G, Soulimane T, Mauk AG (1993) Biochemistry 32:10912
204. Rist W, Jorgensen TJD, Roepstorff P, Bukau B, Mayer MP (2003) J Biol Chem 278:51415
205. Lanman J, Lam TT, Barnes S, Sakalian M, Emmett MR, Marshall AG, Prevelige PE (2003) J Mol Biol 334:1133
206. Lanman J, Lam TT, Emmett MR, Marshall AG, Sakalian M, Prevelige PE (2004) Nat Struct Mol Biol 11:676
207. Nazabal A, Laguerre M, Schmitter JM (2003) J Am Soc Mass Spectrom 14:471
208. Kipping M, Schierhorn A (2003) J Mass Spectrom 38:271

209. Jorgensen TJD, Gardsvoll H, Ploug M, Roepstorff P (2005) J Am Chem Soc 127:2785
210. Jorgensen TJD, Bache N, Roepstorff P, Gardsvoll H, Ploug M (2005) Mol Cell Proteomics 4:1910
211. Charlebois JP, Patrie SM, Kelleher NL (2003) Anal Chem 75:3263
212. Pantazatos D, Kim JS, Klock HE, Stevens RC, Wilson IA, Lesley SA, Woods VL (2004) Proc Natl Acad Sci USA 101:751
213. Klose J (1975) Humangenetik 26:231
214. Ofarrell PH (1975) J Biol Chem 250:4007
215. Klose J, Kobalz U (1995) Electrophoresis 16:1034
216. Klose J (1982) Hoppe Seylers Z Phys Chem 363:1004
217. Klose J, Feller M (1981) Electrophoresis 2:12
218. Challapalli KK, Zabel C, Schuchhardt J, Kaindl AM, Klose J, Herzel H (2004) Electrophoresis 25:3040
219. Davis MT, Beierle J, Bures ET, McGinley MD, Mort J, Robinson JH, Spahr CS, Yu W, Luethy R, Patterson SD (2001) J Chrom B 752:281
220. Wolters DA, Washburn MP, Yates JR (2001) Anal Chem 73:5683
221. Han DK, Eng J, Zhou HL, Aebersold R (2001) Nat Biotechnol 19:946
222. Johnson T, Bergquist J, Ekman R, Nordhoff E, Schurenberg M, Kloppel KD, Muller M, Lehrach H, Gobom J (2001) Anal Chem 73:1670
223. Wetterhall M, Palmblad M, Hakansson P, Markides KE, Bergquist J (2002) J Proteome Res 1:361
224. Tsybin YO, Hakansson P, Wetterhall M, Markides KE, Bergquist J (2002) Eur J Mass Spectrom 8:389
225. Bergquist J (2003) Curr Opin Mol Ther 5:310
226. Smith RD (2000) Nat Biotechnol 18:1041
227. Lipton MS, Pasa-Tolic L, Anderson GA, Anderson DJ, Auberry DL, Battista KR, Daly MJ, Fredrickson J, Hixson KK, Kostandarithes H, Masselon C, Markillie LM, Moore RJ, Romine MF, Shen YF, Stritmatter E, Tolic N, Udseth HR, Venkateswaran A, Wong LK, Zhao R, Smith RD (2002) Proc Natl Acad Sci USA 99:11049
228. Zuberovic A, Ullsten S, Hellman U, Markides KE, Bergquist J (2004) Rapid Commun Mass Spectrom 18:2946
229. Ullsten S, Zuberovic A, Wetterhall M, Hardenborg E, Markides KE, Bergquist J (2004) Electrophoresis 25:2090
230. Thorslund S, Lindberg P, Andren PE, Nikolajeff F, Bergquist J (2005) Electrophoresis 26:4674
231. Liljegren G, Dahlin A, Zettersten C, Bergquist J, Nyholm L (2005) Lab Chip 5:1008
232. Dahlin AP, Bergstrom SK, Andren PE, Markides KE, Bergquist J (2005) Anal Chem 77:5356
233. Zhang KL, Tang H (2003) J Chromatogr B Analyt Technol Biomed Life Sci 783:173
234. Zhang KL, Tang H, Huang L, Blankenship JW, Jones PR, Xiang F, Yau PM, Burlingame AL (2002) Anal Biochem 306:259
235. Zhang KL, Williams KE, Huang L, Yau P, Siino JS, Bradbury EM, Jones PR, Minch MJ, Burlingame AL (2002) Mol Cell Proteomics 1:500
236. Zhang KL, Siino JS, Jones PR, Yau PM, Bradbury EM (2004) Proteomics 4:3765
237. Deterding LJ, Banks GC, Tomer KB, Archer TK (2004) Methods 33:53
238. Zhang X, Guan SH, Chalkley RJ, Recht J, Diaz RL, Allis CD, Marshall AG, Burlingame AL (2006) FASEB J 20:A100
239. Nordhoff E, Krogsdam AM, Jorgensen HF, Kallipolitis BH, Clark BFC, Roepstorff P, Kristiansen K (1999) Nat Biotechnol 17:884

240. Pierro P, Capaccio L, Gadaleta G (1999) FEBS Lett 457:307
241. Yaneva M, Tempst P (2003) Anal Chem 75:6437
242. Ranish JA, Hahn S, Lu Y, Yi EC, Li XJ, Eng J, Aebersold R (2004) Nat Genet 36:707
243. Ranish JA, Yi EC, Leslie DM, Purvine SO, Goodlett DR, Eng J, Aebersold R (2003) Nat Genet 33:349
244. Himeda CL, Ranish JA, Angello JC, Maire P, Aebersold R, Hauschka SD (2004) Mol Cell Biol 24:2132
245. Brand M, Ranish JA, Kummer NT, Hamilton J, Igarashi K, Francastel C, Chi TH, Crabtree GR, Aebersold R, Groudine M (2004) Nat Struct Mol Biol 11:73
246. Mirgorodskaya E, Braeuer C, Fucini P, Lehrach H, Gobom J (2005) Proteomics 5:399
247. Wall DB, Berger SJ, Finch JW, Cohen SA, Richardson K, Chapman R, Drabble D, Brown J, Gostick D (2002) Electrophoresis 23:3193
248. Ericson C, Phung QT, Horn DM, Peters EC, Fitchett JR, Ficarro SB, Salomon AR, Brill LM, Brock A (2003) Anal Chem 75:2309
249. Bodnar WM, Blackburn RK, Krise JM, Moseley MA (2003) J Am Soc Mass Spectrom 14:971
250. Zhen YJ, Xu NF, Richardson B, Becklin R, Savage JR, Blake K, Peltier JM (2004) J Am Soc Mass Spectrom 15:803
251. Lochnit G, Geyer R (2004) Biomed Chromatogr 18:841
252. McDonald C, Li L (2005) Anal Chim Acta 534:3
253. Chen HS, Rejtar T, Andreev V, Moskovets E, Karger BL (2005) Anal Chem 77:2323
254. Ro KW, Liu H, Knapp DR (2006) J Chromatogr A 1111:40
255. Bandeira N, Tang HX, Bafna V, Pevzner P (2004) Anal Chem 76:7221
256. Eisenberg S, Francesconi SC, Civalier C, Walker SS (1990) Methods Enzymol 182:521
257. Gabrielsen OS, Hornes E, Korsnes L, Ruet A, Oyen TB (1989) Nucleic Acids Res 17:6253
258. Gabrielsen OS, Huet J (1993) Methods Enzymol 218:508
259. Gadgil H, Oak SA, Jarrett HW (2001) J Biochem Biophys Methods 49:607
260. Gadgil H, Jurado LA, Jarrett HW (2001) Anal Biochem 290:147
261. Kadonaga JT (1991) Methods Enzymol 208:10
262. Roberts SGE, Green MR (1996) Methods Enzymol 273:110
263. Cheng XH, Morin PE, Harms AC, Bruce JE, BenDavid Y, Smith RD (1996) Anal Biochem 239:35
264. Cheng XH, Harms AC, Goudreau PN, Terwilliger TC, Smith RD (1996) Proc Natl Acad Sci USA 93:7022
265. Tontonoz P, Graves RA, Budavari AI, Erdjumentbromage H, Lui M, Hu E, Tempst P, Spiegelman BM (1994) Nucleic Acids Res 22:5628
266. Winkler GS, Lacomis L, Philip J, Erdjument-Bromage H, Svejstrup JQ, Tempst P (2002) Methods 26:260
267. Jarvik JW, Telmer CA (1998) Annu Rev Genet 32:601
268. Fritze CE, Anderson TR (2000) Methods Enzymol 327:3
269. Bauer A, Kuster B (2003) Eur J Biochem 270:570
270. Ho Y, Gruhler A, Heilbut A, Bader GD, Moore L, Adams SL, Millar A, Taylor P, Bennett K, Boutilier K, Yang LY, Wolting C, Donaldson I, Schandorff S, Shewnarane J, Vo M, Taggart J, Goudreault M, Muskat B, Alfarano C, Dewar D, Lin Z, Michalickova K, Willems AR, Sassi H, Nielsen PA, Rasmussen KJ, Andersen JR, Johansen LE, Hansen LH, Jespersen H, Podtelejnikov A, Nielsen E, Crawford J, Poulsen V, Sorensen BD, Matthiesen J, Hendrickson RC, Gleeson F, Pawson T, Moran MF, Durocher D, Mann M, Hogue CWV, Figeys D, Tyers M (2002) Nature 415:180

271. Rigaut G, Shevchenko A, Rutz B, Wilm M, Mann M, Seraphin B (1999) Nat Biotechnol 17:1030
272. Gould KL, Ren LP, Feoktistova AS, Jennings JL, Link AJ (2004) Methods 33:239
273. Tasto JJ, Carnahan RH, McDonald WH, Gould KL (2001) Yeast 18:657
274. Rohila JS, Chen M, Cerny R, Fromm ME (2004) Plant J 38:172
275. Rohila JS, Chen M, Chen S, Chen J, Cerny R, Dardick C, Canlas P, Xu X, Gribskov M, Kanrar S, Zhu JK, Ronald P, Fromm ME (2006) Plant J 46:1
276. Gingras AC, Aebersold R, Raught B (2005) J Physiol 563:11
277. Bertwistle D, Sugimoto M, Sherr CJ (2004) Mol Cell Biol 24:985
278. Bouwmeester T, Bauch A, Ruffner H, Angrand PO, Bergamini G, Croughton K, Cruciat C, Eberhard D, Gagneur J, Ghidelli S, Hopf C, Huhse B, Mangano R, Michon AM, Schirle M, Schlegl J, Schwab M, Stein MA, Bauer A, Casari G, Drewes G, Gavin AC, Jackson DB, Joberty G, Neubauer G, Rick J, Kuster B, Superti-Furga G (2004) Nat Cell Biol 6:97
279. Brajenovic M, Joberty G, Kuster B, Bouwmeester T, Drewes G (2004) J Biol Chem 279:12804
280. Jeronimo C, Langelier MF, Zeghouf M, Cojocaru M, Bergeron D, Baali D, Forget D, Mnaimneh S, Davierwala AP, Pootoolal J, Chandy M, Canadien V, Beattie BK, Richards DP, Workman JL, Hughes TR, Greenblatt J, Coulombe B (2004) Mol Cell Biol 24:7043
281. Krone JR, Nelson RW, Dogruel D, Williams P, Granzow R (1997) Anal Biochem 244:124
282. Nelson RW, Krone JR (1999) J Mol Recognit 12:77
283. Nelson RW, Krone JR, Jansson O (1997) Anal Chem 69:4363
284. Nelson RW, Krone JR, Jansson O (1997) Anal Chem 69:4369
285. Nelson RW, Jarvik JW, Taillon BE, Tubbs KA (1999) Anal Chem 71:2858
286. Sonksen CP, Nordhoff E, Jansson O, Malmqvist M, Roepstorff P (1998) Anal Chem 70:2731
287. Nelson RW, Nedelkov D, Tubbs KA (2000) Electrophoresis 21:1155
288. Nedelkov D, Tubbs KA, Nelson RW (2002) Proteomics 2:441
289. Nedelkov D, Rasooly A, Nelson RW (2000) Int J Food Microbiol 60:1
290. Nedelkov D, Nelson RW (2003) Trends Biotechnol 21:301
291. Nedelkov D, Nelson RW (2000) Anal Chim Acta 423:1
292. Nedelkov D, Nelson RW (2000) J Mol Recognit 13:140
293. Heck AJR, Krijgsveld J (2004) Expert Rev Proteomics 1:317
294. Flory MR, Gingras AC, Keller A, Lee H, Li XJ, Nesvizhskii A, Ranish J, Ye ML, Zhang H, Aebersold R (2004) Eur J Cell Biol 83:88
295. Flory MR, Griffin TJ, Martin D, Aebersold R (2002) Trends Biotechnol 20:S23
296. Julka S, Regnier F (2004) J Proteome Res 3:350
297. Beynon RJ, Pratt JM (2005) Mol Cell Proteomics 4:857
298. Schneider LV, Hall MR (2005) Drug Discov Today 10:353
299. Kusmierz JJ, Sumrada R, Desiderio DM (1990) Anal Chem 62:2395
300. Krijgsveld J, Ketting RF, Mahmoudi T, Johansen J, Artal-Sanz M, Verrijzer CP, Plasterk RHA, Heck AJR (2003) Nat Biotechnol 21:927
301. Gustavsson N, Greber B, Kreitler T, Himmelbauer H, Lehrach H, Gobom J (2005) Proteomics 5:3563
302. Ong SE, Blagoev B, Kratchmarova I, Kristensen DB, Steen H, Pandey A, Mann M (2002) Mol Cell Proteomics 1:376
303. Foster LJ, de Hoog CL, Mann M (2003) Proc Natl Acad Sci USA 100:5813
304. Ibarrola N, Kalume DE, Gronborg M, Iwahori A, Pandey A (2003) Anal Chem 75:6043

305. Andersen JS, Lam YW, Leung AKL, Ong SE, Lyon CE, Lamond AI, Mann M (2005) Nature 433:77
306. Ibarrola N, Molina H, Iwahori A, Pandey A (2004) J Biol Chem 279:15805
307. Ong SE, Mittler G, Mann M (2004) Nat Methods 1:119
308. Regnier FE, Seeley E, Julka S, Mirzaei H, Sioma C, Riggs L (2003) Chem Res Toxicol 16:1687
309. Ishihama Y, Sato T, Tabata T, Miyamoto N, Sagane K, Nagasu T, Oda Y (2005) Nat Biotechnol 23:617
310. Mirgorodskaya OA, Kozmin YP, Titov MI, Korner R, Sonksen CP, Roepstorff P (2000) Rapid Commun Mass Spectrom 14:1226
311. Bantscheff M, Dumpelfeld B, Kuster B (2004) Rapid Commun Mass Spectrom 18:869
312. Yao XD, Freas A, Ramirez J, Demirev PA, Fenselau C (2001) Anal Chem 73:2836
313. Wang YK, Ma ZX, Quinn DF, Fu EW (2001) Anal Chem 73:3742
314. Qin J, Herring CJ, Zhang XL (1998) Rapid Commun Mass Spectrom 12:209
315. Sun G, Anderson VE (2005) Rapid Commun Mass Spectrom 19:2849
316. Mirgorodskaya E, Wanker E, Otto A, Lehrach H, Gobom J (2005) J Proteome Res 4:2109
317. Gygi SP, Rist B, Gerber SA, Turecek F, Gelb MH, Aebersold R (1999) Nat Biotechnol 17:994
318. Hansen KC, Schmitt-Ulms G, Chalkley RJ, Hirsch J, Baldwin MA, Burlingame AL (2003) Mol Cell Proteomics 2:299
319. Li JX, Steen H, Gygi SP (2003) Mol Cell Proteomics 2:1198
320. Oda Y, Owa T, Sato T, Boucher B, Daniels S, Yamanaka H, Shinohara Y, Yokoi A, Kuromitsu J, Nagasu T (2003) Anal Chem 75:2159
321. Olsen JV, Andersen JR, Nielsen PA, Nielsen ML, Figeys D, Mann M, Wisniewski JR (2004) Mol Cell Proteomics 3:82
322. Nielsen PA, Olsen JV, Podtelejnikov AV, Andersen JR, Mann M, Wisniewski JR (2005) Mol Cell Proteomics 4:402
323. Goodlett DR, Keller A, Watts JD, Newitt R, Yi EC, Purvine S, Eng JK, von Haller P, Aebersold R, Kolker E (2001) Rapid Commun Mass Spectrom 15:1214
324. Munchbach M, Quadroni M, Miotto G, James P (2000) Anal Chem 72:4047
325. Ji JY, Chakraborty A, Geng M, Zhang X, Amini A, Bina M, Regnier F (2000) J Chrom B 745:197
326. Schmidt A (2005) Proteomics 5:826
327. Zhang XL, Jin QK, Carr SA, Annan RS (2002) Rapid Commun Mass Spectrom 16:2325
328. Ross PL, Huang YLN, Marchese JN, Williamson B, Parker K, Hattan S, Khainovski N, Pillai S, Dey S, Daniels S, Purkayastha S, Juhasz P, Martin S, Bartlet-Jones M, He F, Jacobson A, Pappin DJ (2004) Mol Cell Proteomics 3:1154
329. Sechi S, Chait BT (1998) Anal Chem 70:5150
330. Shen M, Guo L, Wallace A, Fitzner J, Eisenman J, Jacobson E, Johnson RS (2003) Mol Cell Proteomics 2:315
331. Sebastiano R, Citterio A, Lapadula M, Righetti PG (2003) Rapid Commun Mass Spectrom 17:2380
332. Pasquarello C, Sanchez JC, Hochstrasser DF, Corthals GL (2004) Rapid Commun Mass Spectrom 18:117
333. Zhou HL, Ranish JA, Watts JD, Aebersold R (2002) Nat Biotechnol 20:512
334. Qiu YC, Sousa EA, Hewick RM, Wang JH (2002) Anal Chem 74:4969
335. Shi Y, Xiang R, Crawford JK, Colangelo CM, Horvath C, Wilkins JA (2004) J Proteome Res 3:104

336. Shi Y, Xiang R, Horvath C, Wilkins JA (2005) J Proteome Res 4:1427
337. Thompson A, Schafer J, Kuhn K, Kienle S, Schwarz J, Schmidt G, Neumann T, Hamon C (2003) Anal Chem 75:4942
338. Wang SH, Regnier FE (2001) J Chromatogr A 924:345
339. Che FY, Fricker LD (2002) Anal Chem 74:3190
340. Mason DE, Liebler DC (2003) J Proteome Res 2:265
341. Lee YH, Han H, Chang SB, Lee SW (2004) Rapid Commun Mass Spectrom 18:3019
342. Hsu JL, Huang SY, Chow NH, Chen SH (2003) Anal Chem 75:6843
343. Brancia FL, Openshaw ME, Kumashiro S (2002) Rapid Commun Mass Spectrom 16:2255
344. Brancia FL, Montgomery H, Tanaka K, Kumashiro S (2004) Anal Chem 76:2748
345. Beardsley RL, Reilly JP (2003) J Proteome Res 2:15
346. Peters EC, Horn DM, Tully DC, Brock A (2001) Rapid Commun Mass Spectrom 15:2387
347. Kuyama H, Watanabe M, Toda C, Ando E, Tanaka K, Nishimura O (2003) Rapid Commun Mass Spectrom 17:1642
348. Goshe MB, Veenstra TD, Panisko EA, Conrads TP, Angell NH, Smith RD (2002) Anal Chem 74:607
349. Qian WJ, Gosche MB, Camp DG, Yu LR, Tang KQ, Smith RD (2003) Anal Chem 75:5441
350. Wells L, Vosseller K, Cole RN, Cronshaw JM, Matunis MJ, Hart GW (2002) Mol Cell Proteomics 1:791
351. Amoresano A, Marino G, Cirulli C, Quemeneur E (2004) Eur J Mass Spectrom 10:401
352. Vosseller K, Hansen KC, Chalkley RJ, Trinidad JC, Wells L, Hart GW, Burlingame AL (2005) Proteomics 5:388
353. Leitner A, Lindner W (2004) J Chromatogr B Analyt Technol Biomed Life Sci 813:1
354. Saghatelian A, Cravatt BF (2005) Nat Chem Biol 1:130
355. Verhelst SHL, Bogyo M (2005) Biotechniques 38:175
356. Andersen JS, Lyon CE, Fox AH, Leung AKL, Lam YW, Steen H, Mann M, Lamond AI (2002) Curr Biol 12:1
357. Coute Y, Burgess JA, Diaz JJ, Chichester C, Lisacek F, Greco A, Sanchez JC (2006) Mass Spectrom Rev 25:215
358. Vollmer M, Horth P, Rozing G, Coute Y, Grimm R, Hochstrasser D, Sanchez JC (2006) J Sep Sci 29:499
359. Staub E, Fiziev P, Rosenthal A, Hinzmann B (2004) Bioessays 26:567
360. Mintz PJ, Patterson SD, Neuwald AF, Spahr CS, Spector DL (1999) EMBO J 18:4308
361. Rout MP, Aitchison JD, Suprapto A, Hjertaas K, Zhao YM, Chait BT (2000) J Cell Biol 148:635
362. Neubauer G, Gottschalk A, Fabrizio P, Seraphin B, Luhrmann R, Mann M (1997) Proc Natl Acad Sci USA 94:385
363. Neubauer G, King A, Rappsilber J, Calvio C, Watson M, Ajuh P, Sleeman J, Lamond A, Mann M (1998) Nat Genet 20:46
364. Rappsilber J, Ryder U, Lamond AI, Mann M (2002) Genome Res 12:1231
365. Zhou ZL, Licklider LJ, Gygi SP, Reed R (2002) Nature 419:182
366. Andersen JS, Wilkinson CJ, Mayor T, Mortensen P, Nigg EA, Mann M (2003) Nature 426:570
367. Taylor SW, Fahy E, Ghosh SS (2003) Trends Biotechnol 21:82
368. Liu HY, Badarinarayana V, Audino DC, Rappsilber J, Mann M, Denis CL (1998) EMBO J 17:1096

369. Phan L, Zhang XL, Asano K, Anderson J, Vornlocher HP, Greenberg JR, Qin J, Hinnebusch AG (1998) Mol Cell Biol 18:4935
370. Asano K, Phan L, Anderson J, Hinnebusch AG (1998) J Biol Chem 273:18573
371. Saleh A, Schieltz D, Ting N, McMahon SB, Litchfield DW, Yates JR, Lees-Miller SP, Cole MD, Brandl CJ (1998) J Biol Chem 273:26559
372. Underhill C, Qutob MS, Yee SP, Torchia J (2000) J Biol Chem 275:40463
373. Edvardsson U, Alexandersson M, von Lowenhielm HB, Nystrom AC, Ljung B, Nilsson F, Dahllof B (1999) Electrophoresis 20:935
374. Freitas MA, Sklenar AR, Parthun MR (2004) J Cell Biochem 92:691
375. Zhang LW, Eugeni EE, Parthun MR, Freitas MA (2003) Chromosoma 112:77
376. Burlingame AL, Zhang X, Chalkley RJ (2005) Methods 36:383
377. Pesavento JJ, Kim YB, Taylor GK, Kelleher NL (2004) J Am Chem Soc 126:3386
378. Smith CM, Gafken PR, Zhang ZL, Gottschling DE, Smith JB, Smith DL (2003) Anal Biochem 316:23
379. Chu FX, Nusinow DA, Chalkely RJ, Plath K, Panning B, Burlingame AL (2006) Mol Cell Proteomics 5:194
380. Garcia BA, Busby SA, Barber CM, Shabanowitz J, Allis CD, Hunt DF (2004) J Proteome Res 3:1219
381. Garcia BA, Busby SA, Shabanowitz J, Hunt DF, Mishra N (2005) J Proteome Res 4:2032
382. Sarg B, Koutzamani E, Helliger W, Rundquist I, Lindner HH (2002) J Biol Chem 277:39195
383. Garcia BA, Barber CM, Hake SB, Ptak C, Turner FB, Busby SA, Shabanowitz J, Moran RG, Allis CD, Hunt DF (2005) Biochemistry 44:13202
384. Fischle W, Wang YM, Allis CD (2003) Nature 425:475
385. Henikoff S (2005) Proc Natl Acad Sci USA 102:5308
386. Cosgrove MS, Wolberger C (2005) Biochem Cell Biol 83:468
387. Kamakaka RT, Biggins S (2005) Genes Dev 19:295
388. Albig W, Trappe R, Kardalinou E, Eick S, Doenecke D (1999) Biol Chem 380:7
389. McKittrick E, Gaften PR, Ahmad K, Henikoff S (2004) Proc Natl Acad Sci USA 101:1525
390. Sarg B, Green A, Soderkvist P, Helliger W, Rundquist I, Lindner HH (2005) FEBS J 272:3673
391. Bonenfant D, Coulot M, Towbin H, Schindler P, van Oostrum J (2006) Mol Cell Proteomics 5:541
392. Chik JK, Schriemer DC (2003) J Mol Biol 334:373
393. Sanders SL, Jennings J, Canutescu A, Link AJ, Weil PA (2002) Mol Cell Biol 22:4723
394. Gavin AC, Bosche M, Krause R, Grandi P, Marzioch M, Bauer A, Schultz J, Rick JM, Michon AM, Cruciat CM, Remor M, Hofert C, Schelder M, Brajenovic M, Ruffner H, Merino A, Klein K, Hudak M, Dickson D, Rudi T, Gnau V, Bauch A, Bastuck S, Huhse B, Leutwein C, Heurtier MA, Copley RR, Edelmann A, Querfurth E, Rybin V, Drewes G, Raida M, Bouwmeester T, Bork P, Seraphin B, Kuster B, Neubauer G, Superti-Furga G (2002) Nature 415:141
395. Uetz P, Giot L, Cagney G, Mansfield TA, Judson RS, Knight JR, Lockshon D, Narayan V, Srinivasan M, Pochart P, Qureshi-Emili A, Li Y, Godwin B, Conover D, Kalbfleisch T, Vijayadamodar G, Yang MJ, Johnston M, Fields S, Rothberg JM (2000) Nature 403:623
396. Uetz P, Hughes RE (2000) Curr Opin Microbiol 3:303
397. Dziembowski A, Seraphin B (2004) FEBS Lett 556:1
398. Gavin AC, Aloy P, Grandi P, Krause R, Boesche M, Marzioch M, Rau C, Jensen LJ, Bastuck S, Dumpelfeld B, Edelmann A, Heurtier MA, Hoffman V, Hoefert C, Klein K,

Hudak M, Michon AM, Schelder M, Schirle M, Remor M, Rudi T, Hooper S, Bauer A, Bouwmeester T, Casari G, Drewes G, Neubauer G, Rick JM, Kuster B, Bork P, Russell RB, Superti-Furga G (2006) Nature 440:631

399. Krogan NJ, Cagney G, Yu HY, Zhong GQ, Guo XH, Ignatchenko A, Li J, Pu SY, Datta N, Tikuisis AP, Punna T, Peregrin-Alvarez JM, Shales M, Zhang X, Davey M, Robinson MD, Paccanaro A, Bray JE, Sheung A, Beattie B, Richards DP, Canadien V, Lalev A, Mena F, Wong P, Starostine A, Canete MM, Vlasblom J, Wu S, Orsi C, Collins SR, Chandran S, Haw R, Rilstone JJ, Gandi K, Thompson NJ, Musso G, St Onge P, Ghanny S, Lam MHY, Butland G, Altaf-Ui AM, Kanaya S, Shilatifard A, O'Shea E, Weissman JS, Ingles CJ, Hughes TR, Parkinson J, Gerstein M, Wodak SJ, Emili A, Greenblatt JF (2006) Nature 440:637

Author Index Volumes 101–104

Author Index Volumes 1–50 see Volume 50
Author Index Volumes 51–100 see Volume 100

Acker, J. P.: Biopreservation of Cells and Engineered Tissues. Vol. 103, pp. 157–187.
Ahn, E. S. see Webster, T. J.: Vol. 103, pp. 275–308.
Andreadis S. T.: Gene-Modified Tissue-Engineered Skin: The Next Generation of Skin Substitutes. Vol. 103, pp. 241–274.

Backendorf, C. see Fischer, D. F.: Vol. 104, pp. 37–64
Beier, V., Mund, C., and *Hoheisel, J. D.*: Monitoring Methylation Changes in Cancer. Vol. 104, pp. 1–11.
Berthiaume, F. see Nahmias, Y.: Vol. 103, pp. 309–329.
Bhatia, S. N. see Tsang, V. L.: Vol. 103, pp. 189–205.
Biener, R. see Goudar, C.: Vol. 101, pp. 99–118.
Bulyk, M. L.: Protein Binding Microarrays for the Characterization of DNA–Protein Interactions. Vol. 104, pp. 65–85.

Chan, C. see Patil, S.: Vol. 102, pp. 139–159.
Chuppa, S. see Konstantinov, K.: Vol. 101, pp. 75–98.

Farid, S. S.: Established Bioprocesses for Producing Antibodies as a Basis for Future Planning. Vol. 101, pp. 1–42.
Field, S., Udalova, I., and *Ragoussis, J.*: Accuracy and Reproducibility of Protein–DNA Microarray Technology. Vol. 104, pp. 87–110.
Fischer, D. F. and *Backendorf, C.*: Identification of Regulatory Elements by Gene Family Footprinting and In Vivo Analysis. Vol. 104, pp. 37–64.
Fisher, R. J. and *Peattie, R. A.*: Controlling Tissue Microenvironments: Biomimetics, Transport Phenomena, and Reacting Systems. Vol. 103, pp. 1–73.
Fisher, R. J. see Peattie, R. A.: Vol. 103, pp. 75–156.

Garlick, J. A.: Engineering Skin to Study Human Disease – Tissue Models for Cancer Biology and Wound Repair. Vol. 103, pp. 207–239.
Goudar, C. see Konstantinov, K.: Vol. 101, pp. 75–98.
Goudar, C., Biener, R., Zhang, C., Michaels, J., Piret, J. and *Konstantinov, K.*: Towards Industrial Application of Quasi Real-Time Metabolic Flux Analysis for Mammalian Cell Culture. Vol. 101, pp. 99–118.

Hoheisel, J. D. see Beier, V.: Vol. 104, pp. 1–11
Holland, T. A. and *Mikos, A. G.*: Review: Biodegradable Polymeric Scaffolds. Improvements in Bone Tissue Engineering through Controlled Drug Delivery. Vol. 102, pp. 161–185.
Hossler, P. see Seth, G.: Vol. 101, pp. 119–164.
Hu, W.-S. see Seth, G.: Vol. 101, pp. 119–164.

Hu, W.-S. see Wlaschin, K. F.: Vol. 101, pp. 43–74.

Jiang, J. see Lu, H. H.: Vol. 102, pp. 91–111.

Kaplan, D. see Velema, J.: Vol. 102, pp. 187–238.
Konstantinov, K., Goudar, C., Ng, M., Meneses, R., Thrift, J., Chuppa, S., Matanguihan, C., Michaels, J. and *Naveh, D.*: The "Push-to-Low" Approach for Optimization of High-Density Perfusion Cultures of Animal Cells. Vol. 101, pp. 75–98.
Konstantinov, K. see Goudar, C.: Vol. 101, pp. 99–118.

Laurencin, C. T. see Nair, L. S.: Vol. 102, pp. 47–90.
Lehrach, H. see Nordhoff, E.: Vol. 104, pp. 111–195
Li, Z. see Patil, S.: Vol. 102, pp. 139–159.
Lu, H. H. and *Jiang, J.*: Interface Tissue Engineering and the Formulation of Multiple-Tissue Systems. Vol. 102, pp. 91–111.

Majka, J. and *Speck, C.*: Analysis of Protein–DNA Interactions Using Surface Plasmon Resonance. Vol. 104, pp. 13–36.
Matanguihan, C. see Konstantinov, K.: Vol. 101, pp. 75–98.
Matsumoto, T. and *Mooney, D. J.*: Cell Instructive Polymers. Vol. 102, pp. 113–137.
Meneses, R. see Konstantinov, K.: Vol. 101, pp. 75–98.
Michaels, J. see Goudar, C.: Vol. 101, pp. 99–118.
Michaels, J. see Konstantinov, K.: Vol. 101, pp. 75–98.
Mikos, A. G. see Holland, T. A.: Vol. 102, pp. 161–185.
Moghe, P. V. see Semler, E. J.: Vol. 102, pp. 1–46.
Mooney, D. J. see Matsumoto, T.: Vol. 102, pp. 113–137.
Mund, C. see Beier, V.: Vol. 104, pp. 1–11

Nair, L. S. and *Laurencin, C. T.*: Polymers as Biomaterials for Tissue Engineering and Controlled Drug Delivery. Vol. 102, pp. 47–90.
Nahmias, Y., Berthiaume, F. and *Yarmush, M. L.*: Integration of Technologies for Hepatic Tissue Engineering. Vol. 103, pp. 309–329.
Naveh, D. see Konstantinov, K.: Vol. 101, pp. 75–98.
Ng, M. see Konstantinov, K.: Vol. 101, pp. 75–98.
Nordhoff, E. and *Lehrach, H.*: Identification and Characterization of DNA-Binding Proteins by Mass Spectrometry. Vol. 104, pp. 111–195.

Patil, S., Li, Z. and *Chan, C.*: Cellular to Tissue Informatics: Approaches to Optimizing Cellular Function of Engineered Tissue. Vol. 102, pp. 139–159.
Peattie, R. A. and *Fisher, R. J.*: Perfusion Effects and Hydrodynamics. Vol. 103, pp. 75–156.
Peattie, R. A. see Fisher, R. J.: Vol. 103, pp. 1–73.
Piret, J. see Goudar, C.: Vol. 101, pp. 99–118.

Ragoussis, J. see Field, S.: Vol. 104, pp. 87–110
Ranucci, C. S. see Semler, E. J.: Vol. 102, pp. 1–46.

Semler, E. J., Ranucci, C. S. and *Moghe, P. V.*: Tissue Assembly Guided via Substrate Biophysics: Applications to Hepatocellular Engineering. Vol. 102, pp. 1–46.
Seth, G., Hossler, P., Yee, J. C., Hu, W.-S.: Engineering Cells for Cell Culture Bioprocessing – Physiological Fundamentals. Vol. 101, pp. 119–164.

Speck, C. see Majka, J.: Vol. 104, pp. 13–36

Thrift, J. see Konstantinov, K.: Vol. 101, pp. 75–98.
Tsang, V. L. and *Bhatia, S. N.*: Fabrication of Three-Dimensional Tissues. Vol. 103, pp. 189–205.

Udalova, I. see Field, S.: Vol. 104, pp. 87–110

Velema, J. and *Kaplan, D.*: Biopolymer-Based Biomaterials as Scaffolds for Tissue Engineering. Vol. 102, pp. 187–238.

Webster, T. J. and *Ahn, E. S.*: Nanostructured Biomaterials for Tissue Engineering Bone. Vol. 103, pp. 275–308.
Wlaschin, K. F. and *Hu, W.-S.*: Fedbatch Culture and Dynamic Nutrient Feeding. Vol. 101, pp. 43–74.

Yarmush, M. L. see Nahmias, Y.: Vol. 103, pp. 309–329.
Yee, J. C. see Seth, G.: Vol. 101, pp. 119–164.

Zhang, C. see Goudar, C.: Vol. 101, pp. 99–118.

Subject Index

Actin 181
S-Adenosylmethionine 162
ADP ribosylation 174
Affinity 88
Aging in mammals 177
Alcohol dehydrogenase gene 43
AlignACE 80
Annexin X 43
Anti-glutathione S-transferase, Alexa Fluor® 488 conjugated 77
Association 23

Bacterial cell extracts 103
BARD1 179
Benzophenone derivatives 139
Betaine 75
BIA 157, 181
– core 13, 16
– evaluation, curve fitting 27
Binding site signatures 66
Biomolecular interaction analysis (BIA) 157, 181
BioProspector 80
Biotin–streptavidin 137
Biotinylation 174
BRCA1-BARD1 179
5-Bromouridine 139
Buffer 21

Caenorhabditis elegans 41, 161
cAMP receptor protein (CRP) 174
Carboxy terminus of Hsp70-interacting protein (CHIP) 168
Carboxydextran matrix, coatings 92
CDTF-1 (calcium dependent transcription factor) 55, 56
CE precursor genes 44
Cell extracts 88, 103
Centrosome 172

Chaperones 170
Chip surface 20
ChIP-chip 67, 89
Chromatin 178
Collision-induced dissociation (CID) 118
Consensus 88
Cooperativity fold α factor 30
Cornified cell envelope (CE) precursors 43, 44
Cornulin 44
CpG island 1
Cross-linking 135
Cross-species conservation 80
Culture-derived isotope tags (CDIT) 162
Cy5-dC-puromycin
α-Cyano-4-hydroxycinnamic acid (CHCA) 117, 122
Cyclic adenosine monophosphate receptor protein (CRP) 173

Detection methods 99
Diabetes 172
Differential phylogenetic footprinting 39
Dihydroxybenzoic acid (DHB), glycosylated peptides/proteins 117
Diphtheria toxin repressor 180
Dissociation 24
Dissociation constant Kd 25
DNA binding site motif 65, 80
DNA binding specificity 65
–, proteins 76
DNA methylation 1
DNA methyltransferase-2 (Dnmt2) 113
DNA microarray, quality 75
DNA printing 94
DNA purification 75
DNA sequence, length 93
DnaA 13
DNA-binding domain 88

DNA-binding proteins 111
DNA–protein interactions 65, 88
DNAs, double-stranded 72
Drosophila melanogaster 161
–, homeodomain proteins
  even-skipped/fushi-tarazu 67
–, TF Extradenticle
Duplexes, spotting 93

Electron capture dissociation (ECD) 119
Electron transfer dissociation (ETD) 119
Electrospray ionization (ESI) 113
EMSA-derived binding affinities 92
End-point affinity reaction conditions 99
Endoproteases 163
Epidermal differentiation complex (EDC) 43
Epidermal growth factor (EGF) 162
Epigenetics 1
Equilibrium 25, 99
ESI MS 111
Expressed sequence tag 129

Ferric uptake regulation factor Fur 179
Footprints 38
Foreskin keratinocytes 52
Functional category enrichment 81
Fused genes 44

G-actin 181
GAAP-1 51
Gene expression datasets 81
Gene family 37
– footprinting 41
Genome-wide location analysis 67
Globin 38
Green fluorescent protein (GFP) 71

Helix-turn-helix protein 180
Hexaethylene glycol (HEG) 89
Hill's cooperativity factor 28
Histone deacetylase inhibitor trichostatin A 176
Histones 132, 169, 174
Homeodomain proteins 67
Hornerin 44
Human nucleolar proteins 171
Hybridization 88
Hybridization-based assay 5

Hydrogen/deuterium exchange 141
3-Hydroxypicolinic acid, nucleic acids 117
N-Hydroxysuccinimide esters, coatings 92

Immobilization, dsDNAs 74
Immunoaffinity purification 154
Infrared multiple photon dissociation (IRMPD) 118
Injections 20
Insulin resistance 172
Interactome 182
Intermediate filament associated proteins (IFAP) 44
Involucrin 43, 44
5-Iodouridine 139
Ion fragmentation 114
Ionization 114
Isobaric tag for relative and absolute quantification (iTRAQ) 164
Isotope labeling 158
Isotope-coded affinity tag (ICAT)

Keratinocytes 37, 52
Keratins 41
Kinases/phosphatases 170

Ligands 18
Loricrin 43
Luciferase 67
Lys-C 128

MALDI 111
MALDI-TOF 113
Mammalian cell extracts 104
Mapping, peptide mass 124
Masliner 78
Mass spectrometry 111, 123, 145, 151
Mass transport limitations 19
MDscan 80
MEME 80
Metabolic stable isotope labeling 161
Methylation patterns, microarrays 4
Microarray data quality control 75
Microarray probing 97
Microarray signal intensities 77
Microarray surface chemistry 92
Microarrays, preparation 92
–, types 89

Subject Index

MS   111, 123, 145, 151
–, affinity purification   151
–, biomolecular interaction analysis   157
–, LC/electrophoresis   145
Multidimensional protein identification technology   149
Multiple protein binding sites   27
Murine homeodomain proteins Hmx1 and Nkx2.5   67

Noncovalent complexes   133
Normalisation   102
Nucleobases   139

Oligonucleotide microarray   1

Paralogous genes   41
Peptide–DNA   138
Peptide mass fingerprinting (PMF)   124
Peptide sequence tag   127
Peroxisome proliferator-activated receptor (PPAR)   172
Phosphopeptides   133
Phosphotyrosine proteins   162
Photodamage/photocleavage   139
Poly-d(IC)   152
Poly(ADP-ribose) polymerase-1 (PARP-1)   125, 130
Polyamides   83
Polypeptide–oligonucleotide complexes   140
Post-source decay (PSD)   118
Posttranslational modifications   113, 174
Primer extension   1, 8
Printing buffer   75
Pro-apoptotic nuclear factor GAAP-1   51
Profilaggrin   44
Promoter   37
–, *SPRR3*-specific regulatory elements   51
Protein binding microarrays   65
Protein expression/purification   95
Protein footprinting   135
Protein identification, sequence databases   124
Protein kinase Tra1p   172
Protein modification analysis   129
Protein purification   88
Protein sequencing   128
Protein V   179
Protein-binding microarray   88

Protein–DNA interactions   138
–, BIAcore   18
Psoralen   139
PU.1 transcription factor   179
Purification methods   96

Rate constants *k*on and *k*off   25
Regulatory elements, evolution   38
Repetin   44, 45
Restriction enzymes   89
Retinoid X receptor alpha (RXRα)   174
Ribosomes   134
RNA helicases   170
RNA polymerases   170
RNAi   156

Saccharomyces cerevisiae   68
SAGE-SELEX   69
Sarkosyl   75
Self-hairpinning   72
Sensitivity   97
Sequence selection   92
Shotgun proteomics   150, 164
SILAC   162
Sinapic acid (SA), large peptides/proteins   117
Slide blocking   95
Spliceosome   172
Spot quality   75
SPR   13
SPRR3 promoter   50
Stable isotope labeling   111
Streptavidin   66
Subproteomes   150
Sumoylation   174
Surface labeling   135
Surface plasmon resonance (SPR)   15, 157
Surface-enhanced laser desorption/ionization (SELDI)   122
Systemic lupus erythematosus (SLE)   176

Tandem affinity purification (TAP)   155
Tandem MS (MS/MS)   114
TATA box binding transcription factor TFIID   181
Terminal differentiation   37
4-Thiouridine   139
Time-of-flight (TOF)   114

Tobacco etch virus (TEV) protease   155
Transcription factor   37, 66, 80, 88
– binding sites   38
Transcription initiation complex   151
Transfection   37
Trichohyalin   44, 45
Trypsin   128

Ubiquitylation   174

V8 protease   128

Wash stringencies   101

Zinc fingers   77, 82